. . . .

EUCLID'S
ORCHARD

═══ & ═══

OTHER ESSAYS

. . . .

EUCLID'S
ORCHARD

═══ & ═══

OTHER ESSAYS

THERESA KISHKAN

MOTHER TONGUE PUBLISHING LIMITED
290 Fulford-Ganges Road, Salt Spring Island, B.C. v8κ 2κ6 Canada
www.mothertonguepublishing.com
Represented in North America by Heritage Group Distribution.

Book Design by Setareh Ashrafologhalai
Cover photo: Shutterstock
Typefaces: RTF Amethyst Pro and Sero Pro
Printed on Enviro Cream, 100% recycled
Printed and bound in Canada.

Mother Tongue Publishing gratefully acknowledges the assistance of
the Province of British Columbia through the B.C. Arts Council and we
acknowledge the support of the Canada Council for the Arts, which last
year invested $153 million in writing and publishing throughout Canada.

Nous remercions de son soutien le Conseil des Arts du Canada, qui a
investi 157$ millions de dollars l'an dernier dans les lettres et l'édition à
travers le Canada.

Library and Archives Canada Cataloguing in Publication

Kishkan, Theresa, 1955-, author
     Euclid's orchard and other essays / Theresa Kishkan.

ISBN 978-1-896949-63-5 (softcover)

     I. Title.

PS8571.I75E93 2017   c814'.54   C2017-903550-9

*For my grandchildren, Kelly, Arthur, and Henry Pass,
and in memory of my parents, Shirley and Anthony Kishkan*

(How many generations of dragonflies and monarch butterflies?
How many generations of children ...)

. . . .

TABLE OF CONTENTS

═══════════

HERAKLEITOS ON THE YALAKOM

"How can you hide from what never goes away?" HERAKLEITOS

There were mornings when you went out of the tent alone to cast your line from a dock, the bank of a river, from your small rowboat in the middle of a lake. There were mornings when the campfire snapped and smoked, and you did too, frying fish to have with pancakes, a cigarette propped in a tuna can. We were afraid of you. When you lost your temper, you'd strike out with your hands, you'd take the basement stairs two at a time, and raise your fists to the son you'd felt insulted by, who'd talked back, who'd refused a chore. You rarely listened and seldom praised.

But you knew fish. Had a tackle box filled with lures and sinkers, a ruler to measure your catch, even a small brass scale with a single hook to hang a fish from its gills for an accurate weight. You gave a rod to anyone who seemed interested—each of my children has one; each has a tackle box. You brought old hockey sticks home from the rink where my brother's team practiced and built campstools from the strong maple, canvas seats sewn on with thick black thread. We sat on these while waiting for our pancakes, the crispy fillets of fish. You'd always say what kind of fish you'd caught. Bass, smallmouth and largemouth; trout, rainbows, cutthroat: beauty brought up from the lake on a hook. You didn't know a terrible thing was happening to your son; when you found out, you shrugged. What could you do? That boy, now in his fifties, still waits for you to tell him you're sorry it happened to him. But perhaps you don't care. No, that's not fair. But the knot of your caring is tied up so tightly that there's no way to figure out how the threads of emotion come together to make it hold. And you taught knots to Scouts and know all the best ones—which can be relied on and which can't. A reef or a granny, a half-hitch or Turk's head.

You hated to be asked for anything—a ride to or home from an event, money, anything that might take you out of your regular routine. Which you also hated. You sighed in the morning as you got ready for your job. Sighed as you contemplated the weekend mowing of the grass. Sighed if our mother had to go away and you had to cook. Yet you'd fix some acquaintance's radio or build a gunstock from scratch, taking hours to shape the beautifully grained wood, then sanding it, polishing it. We'd hear the radio come to life in the basement in a cloud of cigarette smoke, and sometimes you hummed while you worked at your bench. And when you did cook, we loved the meals. You'd announce the menu like a chef in a fancy restaurant: Meatloaf Wellington with creamed potatoes; Soulfood, which I remember as ham hocks with barley; once, a split-pea soup, rich and smoky, that I've never forgotten.

My knots were clumsy. Sometimes you demonstrated patience, your hands over mine to guide them, the rope rough against my fingers. And sometimes you told stories to go with the knots. The midshipman's hitch, a knot to use for lines under tension—a vessel in a storm, tied to a pier; or your own tent, on a hunting trip, for big horn sheep in the mountains above the Yalakom River. You took your fishing rod on those trips too, and I imagine you casting into those rivers, the line echoing the current, the smell of trout in a pan if you were lucky.

You kept your knives sharp. Occasionally I'd sit on the basement stairs and hear you running them along a whetstone kept in a wooden box. I loved the smell of the oil, which I inhaled like perfume. You had a collection of hunting knives, each with its special sheath you made. My brothers received knives for gifts, but they never took care of them the way you did. I remember a thin knife for cleaning fish right on the spot so the guts could be tossed back to the water. And another knife, very sharp, with a blade suitable for filleting and a serrated part for taking off a head or tail. It could cut through thick bone. It has a homemade handle of some kind of antler, shaped to fit the hand, and neatly rivetted. You gave it to my second son, thinking him a kindred soul. The thing is, he isn't. He doesn't care about fishing (you told me you'd like him to have your inflatable boat) but

Old fishing knife

cares about getting along—when you talk, he listens, and you've mistaken listening for a shared passion for fishing. (He won't eat fish.) I always wanted to be taken seriously, wanted to learn fly-fishing, but I also wanted to defend my ground, say my piece, and you couldn't bear anyone talking when you thought yours should be the only voice in the conversation. My son doesn't care about the knife, he left it on a shelf, but I care about it and have it on my desk so I can run my thumb along the rough chalky end of the antler handle and think about the places the knife might have been—the Cowichan River; lakes strung like stars up the spine of Vancouver Island, one even called Stella;

rivers and lakes in the north; rarely the chuck, but even the lakes near my home on the Sechelt Peninsula where you'd take my children out in your inflatable boat and teach the patience of the hook. The patience must have come with age because I remember only your temper, your irritation at being asked for something, the bitter words about ingratitude. Yet they sat with you for hours and thought you a perfect grandfather.

And how, no, *why*, do I imagine things could have been different? You were the youngest child in a family who'd seen terrible hardship. Like so many families in Canada, yours had its origins in Eastern Europe. Your mother lost her first husband, the father of her first nine children, one of whom died in infancy. Her second marriage produced two children, your only full sister, Julia, dying three years before your birth. Your parents barely spoke English. You said your mother attended school with you when you were six so she could learn to write, her large body somehow fitting into the chairs in a primary classroom. Of course this brings me to tears. Your parents were struggling to make a living so you were raised mostly by your grown half-sisters. They adored you, their youngest brother Tony, gave you every attention, and made you into one of those boys convinced of their superior authority.

In the basement, you once showed me to how aim a rifle. I remember the intense focus, the pleasure of finding the target through the sight, and for a moment I almost understood your obsession with guns. You had a collection of them, which you made parts for and kept in good repair. But I couldn't understand the next step—taking the life of an animal, even though I ate meat and knew all those arguments about respecting what one consumed. The way you mocked me for this made it difficult to try to find words for what I believed. A teenage girl doesn't necessarily have the language for paradox. Later, reading Herakleitos, I thought I understood something about you and something about me. Our river was never the same. How could it be? I still hear the sneer in your voice as I struggle for the right way to say things. Your children all know that sneer. But your grandchildren don't. By the time they arrived, you'd softened, allowed yourself to be gentle with children finding their way in the world.

At this point in my life, I am trying to make sense of the knotwork of relationships—yours to me, mine to you and my mother, the loops my brothers make in and amongst you, me, how we are tied to each other, places, hitched and slipped and reefed. The beauty of net-making or macramé, intricate and practical in its application to hammocks, plant holders, screens to keep things hidden or sheltered; the bewildering tangle of cat's cradles or rats' nests that my own attempts at macramé resembled. I wanted to make something that would elicit your praise. But mostly you noticed your sons. What you expected of them and how they couldn't not measure up, on the ice rink, the soccer field, the baseball diamond where you coached from the sidelines, irritated beyond belief by mistakes or fumbles. I don't recall a single instance of you giving me the gift of how my own future might unfold, what opportunity might exist for me. Yet my brothers might say the same thing, the future you urged them towards not of their own choosing. I found other adults—teachers, mentors, lovers—who told me how competent I was, how intelligent, how filled with promise. I carried their praise around like something fragile and not quite earned but treasured nonetheless. I wanted to hear it at home, from you, and always suspected I was less than others suggested I was. If I had talent, or gifts, wouldn't you have noticed, and told me so? Yet the paradox is that the waters of your river are also mine.

I try to understand why you would sit in the nice restaurant, martini half-finished, and comment, when my husband said something sweet about me, "I only have one daughter and she hates me." Your mouth small and bitter, your eyes cold. Or why you would say about a grandchild at university, "She doesn't have the brains to do medicine." That child's father will no longer speak to you (the one who talked back all those years ago and reduced you to rage), in part because of your legacy of diminishment. They came so easily off your lips—those phrases that sliced into one's confidence like a sharp knife into a trout.

I don't hate you. You are my father, for better and for worse. And it's getting worse. You are barely civil in restaurants, the servers (invariably women) recipients of your contempt: your martini too warm, your pasta not what you wanted (you order pasta all the time though you insist you hate it), the sauce too spicy, the wine too expensive.

You gave me a dirty metal plate with two photographs printed on it: my grandmother and the child born a few years before you, a sister you never knew because she died as an infant. You acted like you were giving me a precious treasure, and in a way you were—there is so little physical evidence of your family, so few photographs, almost no stories (I will return to this), several letters addressed to you at sea—but this plate was filthy, needed a good wipe, had been kept in the basement with all the detritus of a life. (You have one more photograph, you and your parents, and that same image of the sister inserted just behind your shoulders, hanging in air. The baby floating in the background had been dead for five or six years when the picture was created. Your parents had so little money but wanted a family portrait. I would have thought you would cherish this picture, framed and ready to hang, but it ended up in the closet of a room downstairs, covered in dust. I asked to have it, but you were offended. "If I wanted you to have it, I'd give it to you. It's *mine*." And then you waxed briefly sentimental about your parents, but you never cleaned or hung the portrait.) Everything knotted and tangled and resistant to attempts to sort out the fibers.

When I asked about your father and mother, their relatives, family in Canada and in their native terrain (I can't say countries because they both came from areas portioned and reportioned out at various times in history), you said, "I never thought to ask." You said it as though it surprised you that I might be interested in their lives, that I might be interested in people I am related to over time and generations. Their exact beginnings, their connections to a place, their language and history—all my life I've wanted to know something of this. My oldest son has attempted to locate relatives of your father, pouring over online databases and family trees posted on websites. But when he tells you about them, you are skeptical and suspicious, anticipating a ne'er-do-well on the front step, wanting something from you. And I wonder (but don't ask) what in your past might have suggested to you that relatives only want to take and not give. I remember aunts preparing feasts when you returned to Edmonton, giving us presents, taking us into the heart of their activity. My son has found a man in Ukraine who might be your cousin's son and who has written interesting emails to me to tell of his father and grandfather and their terrible ordeals under Stalin's imposed hunger. He works for an agency that does environmental assessments, and he's passionately interested in his country's history. But when I mentioned him to you, you sneered and said, Just wait. He'll want money or else he's looking for an easy way to get to Canada.

At Christmas, the ghosts hover in my heart and mind. Some years you've come to share our holiday, and there are moments when you've looked genuinely happy. Sitting in a chair by our fire with a drink of eggnog or a glass of wine, you've smiled and remembered Christmases past. You watched *A Christmas Carol* with Alistair Sim and greeted each spirit arriving on screen with a quoted line or two of dialogue, a comment about redemption. But this year, you stay home. Which is the new apartment you've moved to with my mother. You have a view of the city below you and a small mountain behind you—deer browse the shrubs outside the laundry room, and hummingbirds dart at your windows all year round, even at Christmas, small visitors in seasonal colours. It's a good thing you have them to watch, you told us, as you've observed that none of your family will come over the holiday. We all have busy lives, particularly at this time of year when our own children come from afar, and none of us lives in proximity to your city. The three of us who still contact you have all encouraged you and our mother to move closer to us so that we can help with shopping, doctors' appointments, and so forth, but you (understandably) cling to the conviction that you need to live in the city you retired to even though almost all your friends have passed away. You hate the ferries. You hate the traffic in Vancouver that needs to be negotiated if you visit us.

I see a pattern if I look hard enough. Make plans through the winter for trips that might please my mother—to Halifax where she grew up and where one friend still lives; to Salt Spring Island where each of you has happy memories. Maybe to us. Maybe to my brother who lives in the northern part of the province. But then by spring, or summer, nothing has improved, so how could you possibly travel? You can barely walk. Not that you will do anything that might make things better: hydrotherapy (you don't want to appear in a swimsuit); mall-walking even with the many seniors who lace on their running shoes (or fasten them with Velcro, determined to do what they can to remain active) and walk through the big warm shopping malls before the stores open in the morning, staying limber and having fun while they do it. When you were hospitalized once after a fall and when your physical assessment showed that you had hardly any mobility at all, you went twice daily for a course of physiotherapy before they'd let you return home. You hated it. But managed to move a little more easily (though you'd never admit it) in order to make your discharge possible. Our mother said she couldn't cope with you at home unless you could stand up on your own and get yourself to the bathroom, and you were furious enough at her to stay the course. But once home, you stopped doing anything. "What's the point," you'd say. "Doesn't work."

You. Yes, you. I'm talking to you. But typically you don't listen. Can't hear. And now you're dead. I wasn't even there, in Victoria, standing by your bed, but in Venice, eating delicious meals and drinking a glass of Prosecco each afternoon as my husband and I returned from our explorations of that city's shabby extraordinary beauty. I saw you earlier in the fall, but you couldn't have cared less. You asked about my brother, the one that had cut himself off from you completely. ("What do you hear from him," you wondered. And I didn't say that he had severed his knot with a sharp knife, not bothering to untangle and trace the strands back to their origins.) The other two brothers came to your city when you died, helping our mother, organizing paperwork, tying up the loose ends of a life knotted beyond my own understanding, though I will try. I will learn the knots themselves and how to tie them and untie them, not resorting to the knife you left us. The bowline, that creates a temporary (as opposed to a permanent) eye splice, non-slipping loop in the end of a line. The round turn and two half-hitches, a constrictor knot, meaning the tighter you pull on the line, the tighter the knot gets. Also, it is one of the few knots that can be tied or untied with tension in the line. And the rolling hitch, a knot used to take the strain off another line or object.

We wanted to take you for one last camping trip the summer before you died. We planned it anticipating the difficulties. We would sleep in a tent but found a cabin for you and Mum. There could be a campfire, even a dock to fish from. We could make buckwheat pancakes in an old cast-iron skillet the way you liked them. It wasn't too far from your apartment, so we wouldn't need to drive for long. But you wouldn't consider it, though Mum tried to coax you. If it couldn't be the old way, the tent tied to the ground with a tautline hitch, your boat on the shore, tackle box stocked with barbless hooks, the scale waiting to offer its verdict, air mattresses leaning on a woodpile to dry, then you didn't want any part of it. What's the use, you said; it wouldn't be the same. The old-fashioned knots you'd tied long ago wouldn't loosen enough to let you take ease in a chair on the deck of a cabin. And by then the fish knife had gone to my son. Herakleitos on the Yalakom River, on the Cowichan, on the far-seeing MacKenzie where you were young, the Red Deer, all those waters changing as we changed—and were ever the same. Always roads leading to them, and away.

One of your last requests was for the brother who wouldn't come.

NOTE

The passages of Herakleitos are from *Herakleitos and Diogenes*, Guy Davenport (trans.). Bolinas, California: Grey Fox Press, 1979.

My mother's camphorwood chest, with My Sin perfume and assorted papers, 2016

. . . .

TOKENS

===

"Hearts appear cut and drawn on paper, fabric and
parchment and their symbolism is obvious ... Many of
the fabric pieces are taken from sleeves."[1]

1.

We visited the Foundling Museum in early February of 2012, a
year and three months after my mother, Shirley Kishkan, died.
We'd stopped in London for a few days on our way to Brno to
teach a week-long course at Masaryk University, and we wanted
to explore Bloomsbury, near our hotel, on that particular day.
The Foundling Museum provides a glimpse into the culture of
London's Foundling Hospital, founded in the mid-18th century
as a place for unwanted babies or children their parents could
not care for. (I am learning that these are two different concepts.)
The Museum resides in a gracious building on Brunswick Square,
near Coram's Fields, named for the founder of the Hospital. The
Fields are a place for children; any person over the age of six-
teen is considered an adult and must be accompanied by a child

to enter the Fields with their sandpits and football pitches, their swings and duck pond, a petting zoo, and trees for dreaming under.

My mother was a foundling. From *The Concise Oxford Dictionary*: n. Deserted infant of unknown parents. She was born to an unwed mother and given to a foster mother to raise until adoptive parents could be found. But the foster mother never released her, nor adopted her. I have puzzled over this for a long time.

Her foster mother made and kept the distinction between my mother and her two biological children. Their father, a doctor, died after the Halifax explosion, not from injury but from succumbing to the Spanish influenza, due—we were told—to exhaustion from having helped so many victims of the explosion. One story my mother told: Emma Watson, her foster mother would go to the homes of people who wanted to adopt Mum and would return saying, "I can't offer much, but I can offer as much as they can." She had a kind of pride, a stubborn resistance to letting my mother go. But was it love? In the late 1920s and early 1930s, a child in my mother's circumstances was marked. How simple it would have been to erase some of that mark by adopting her, giving her a name with a legacy of stories to nestle into.

My mother had her biological father's name: MacDonald. She knew her biological mother's name: MacDougall. She was born in Sydney, on Cape Breton Island. Her foster mother was in Halifax. Why didn't she stay in Cape Breton, why wasn't she fostered there, or adopted there? I wondered if there was some connection between her biological parents and her foster mother. She was told her biological father was the brother of a Halifax doctor. But no more than this.

At the Foundling Museum, I was fascinated by the cases of tokens left with children accepted at the Foundling Hospital by their parents, usually their mother. Tiny keys, buttons, hairpins, a spyglass, thimbles, coins, playing cards, a single hazelnut, a bone fish. There were small fragments of fabric, snipped from an item of clothing, a piece of patchwork with half a heart stitched on it in red. All of these were carefully wrapped with billets—forms containing information about the child: serial number, date of admission, physical characteristics—and then sealed with wax, the child's name and number written on the outside of the packet. If a parent returned, he or she would have to describe the token or bring a matching fragment of cloth. By then the child's name would have been changed in order for a new life to begin. Charles Bender became Benjamin Twirl, but when his mother presented the other half of the red heart on patchwork, he re-entered his old life. Or as was often the case, the child had already died. Two-thirds of the children admitted to the Hospital died, a reflection of the high infant mortality rate in 18th century cities where poverty, poor sanitation, and a lack of general medical care were endemic.

2.

"The most important fact about the tokens is that ... they were left as *identifiers*—they were not gifts for the children, keepsakes or love tokens, as has often been stated. They were official 'documents'—easily recognizable items that could be used to prove the identity of an infant if the parent or parents found themselves in circumstances to take it back."

After my mother's death, I took so many papers and photographs home, gradually sorting them in a cursory way to figure out how and where to store them. One bag contained items sent to my mother, Shirley, by a neighbour of her foster mother, Emma Watson, and foster sister, Helen Watson, in Halifax. The neighbour had been helpful to both women and inherited the contents of their house after Helen's death. My mother received a thousand dollars and a charm bracelet; recent research on my part reveals that her foster sister's estate was a million dollars, one-third of which went to her church. The bag contained a photo album with pictures of my mother as a child. She'd never had anything like this to show us when we were growing up and asked her about her childhood, so it was a surprise to see "little Shirley with Toby," "Shirley with Bobby"—my mum was Shirley MacDonald in those days—and many papers of various sorts: her foster sister's nursing history, clippings about Halifax history, and to my surprise, a Mylar envelope with the remnants of a telegram in it.

The telegram is foxed and tattered; signs indicate that it was once folded, and in fact the top is still folded, the part indicating that the form is the Canadian Pacific Railway Company's Telegraph and outlining the terms and conditions. I mention the folds because it looks like someone took the time to try to straighten and preserve the telegram, attention not given to other papers in the materials received from Helen Watson's neighbour.

The telegram was sent from Glace Bay, N.S. on December 9th, 19__, the year unreadable, to Dr. Watson, 108 Agricola Street, Halifax, and the message reads as follows: Can you take my practise at once and remain four weeks. I am to be operated on tomorrow for appendicitis. Answer and state terms. Then the name of the sender: the initial(s) is/are hazy and occur just where the telegram has been folded, but the surname is MacDonald.

Dr. D.T.C. Watson would have been my mother's foster father if he had lived, but he died eight years before her birth, eight years before Emma Watson took a foundling into her home. He was born in Jamaica to a family of Scottish sugar traders from Mull; the family moved back and forth as the business flourished and changed. It's unclear how he ended up in Halifax, but Helen told my father he'd been part of the Grenfell Mission, providing medical services to remote communities in northern Newfoundland and Labrador. My father remembered a journal from Dr. Watson's Mission years kept in a rolltop desk in the Watson household that he said should have been part of the National Archives, and I've often wondered what happened to it after my mother's foster sister died.

I send out emails to the library in Glace Bay, to Vital Statistics in Halifax, and to a Cape Breton genealogy site, hoping that someone will be able to tell me something about my mother's biological parents, something about Dr. MacDonald who sent the urgent telegram and who I hope might prove to be a link in the chain of my mother's family. I study the telegram for hidden meaning and come up with nothing. Or everything: "official 'documents'—easily recognisable items that could be used to prove the identity of an infant … " Is this why the telegram was saved and kept with a photo album showing my mother as a curly-haired child called Shirley, in a garden with dogs?

Most emails go unanswered, apart from a reply from Vital Statistics. Rules are clear, I'm told: birth certificates and records are restricted documents I cannot access until a hundred years after my mother's birth. That will be 2026.

Will I still be alive in 2026? Will I remember? Will the trail, however faint it might be now, however overgrown and forgotten by almost everyone alive, have disappeared completely? I put my trust in the agencies on Cape Breton Island itself, hoping that they will recognize that "the most important fact about the tokens is that … they were left as *identifiers*," and that they help me to make an identification, across eight decades and a

My mum, Shirley, right, with unknown child and dog, circa 1930

continent. We cast my mother's ashes into water on the west coast of Vancouver Island as well as under a tree in my garden. She will never know the sad circumstances of her birth, but I want to add to the archive I am keeping in my heart, made of small details: that her hair was curly in childhood, that she wanted a middle name (she loved the name Sybille), that the Halifax she grew up in was full of old houses and leafy streets, that someone heard her first infant cry and either reached for her with love, or didn't. Or that a foxed telegram might be a token, or not. That somewhere a small scrap of fabric waits, cut from a sleeve.

---

3. The Widow's House

As an adult, I seldom asked her what she knew about her biological parents, though I did try, at least twice. The first time, she cried. The second time, she said her foster mother had discouraged her from trying to find them, saying that they knew where she was and never contacted her. My mother told me that she had decided to figure out who they were when our family went to Nova Scotia in 1963, but seeing her foster mother after a long absence—my parents lived in Halifax in 1953 before moving to the West Coast—convinced her that this was her mother, this was the person who'd raised her and to whom she owed loyalty and love, and she abandoned her plan to locate her birth parents. Could I try, I asked. And she was fierce in her disapproval.

So I quietly put the notion aside. But she did say that her foster mother had a copy of her birth certificate with the names of both parents. Her birth mother was a MacDougall and her father, a MacDonald. And his was the surname she had until she married my father in 1950. She'd been told that her biological father was the brother of a prominent Halifax physician.

Her foster mother wanted to control the lives of those around her. Perhaps this was out of fear—she didn't want to lose what she had so carefully created in her widow's house: a garden with little girls in dresses smiling for the camera, a few small dogs, a daughter who was a nurse and who could care for her in her old age. Her iron sense of what was right ruled the household. Right, but also selfish: her daughter Helen was courted by young men, at least one of whom she wanted to marry, but the matriarch soon put a stop to that. The foundling might be less grateful if she made contact with her birth parents and discovered that they had regrets, that they wanted a way to include her in their lives, their families. Or a child completely convinced of its abandonment might want true affection. The stamp of official adoption to seal her love.

Later, when I called the agency responsible for adoption in Nova Scotia, a kind woman told me that adoptee records are accessible. "Are you sure your mother wasn't formally adopted?" she asked, hoping to help. "Because then we could acquire her birth certificate." But no, she was never adopted. And foster children had no such privilege. Nor did their descendents.

4. Tweed

"To us, a small scrap of fabric linking mother and child seems ineffably fragile…"

I was in grade one the year my father's naval ship went to Asia for three months, and for that period, my mother cared for us alone. When I think about it, I can't imagine it was entirely unexpected—she'd married a sailor after all, and he'd been in the navy ever since she'd known him. We'd lived in Matsqui for four years of my early childhood, before we moved back to Victoria, the city of my birth, in the summer of 1962, and for those years, my father worked at the radar base just behind the family housing where we lived in the house at the end of the boulevard, next to Sward's farm. He went to work each morning and returned each evening. So this might have been the first time he'd been away for such a long time.

Four children. She didn't drive. There was a big grocery store, a Safeway, I think, on Fort Street, near Douglas, and some days after school, she'd take my younger brother and me with her to do the shopping. We walked. We almost always walked in those years, though buses ran down Cook Street and maybe even along May Street. We were not well-off, and I'm sure she worried about the fares, so we walked. How far would it have been? I consult a map and count: 24 blocks. Each way. And for the walk back, we all carried two bags. This was before plastic bags with handles, so imagine my brother and me with our paper bag holding macaroni, dog food, carrots, and tins of soup. Some days my mother bought us a treat—Lifesavers or a stick of gum.

I remember how she sat by a window with a cup of instant coffee, a cigarette, and a faraway look in her eyes. I knew even then that she longed for his return.

And one day I returned from school to find her in her—*their*—bedroom in a new suit and coat, the suit powder-blue with dark velvet trim on the pockets and collar. The skirt skimmed over her hips and flared a little, mid-calf. She looked so elegant, hardly like my mother at all, in seamed stockings and small blue shoes. The coat, she told me reverently, was Harris Tweed. It had come from W&J Wilson on Government Street, by Trounce Alley. It was wool, its colours soft and muted: grey, brown, an aqua thread that somehow reminded me of birds. "I'll have it for my whole life, so it's worth every penny!" She said this with such vehemence that I knew she felt guilty about how much she had spent. She let me smell it, and that scent imprinted in me: earthy, not unlike animals, and of her somehow too.

The day my father returned, we met the ship at CFB Esquimalt, familiar ground in some ways. My brothers and I took swimming lessons there, at the Naden pool, free for children of naval personnel. But this time we waited on a dock, or some sort of jetty, while men gathered on the deck of the ship and waved to their families. I can't remember seeing my dad, but of course he was among them, and we went aboard the ship for a tour and refreshments and to see where he slept all those months he was away. It was not luxurious. And he was strange to us, or to me at least, in his naval uniform, among men with whom he'd drunk rum daily, maybe never using his Old Spice on Sunday mornings as he did at home, or polishing his shoes before church.

And he brought presents, packed into a fragrant carved chest, which was my mother's main gift, along with a bottle of My Sin, bought (I'm supposing) at a duty-free shop. I received two cotton dresses, identical madras but for the colours: one blues and greens, and the other rusts and browns. Buttons down the bodices, generously full skirts. I loved them. When I twirled down the sidewalk on my way to school, I felt like a dancer. There was also a little music box for jewellery. I still have it, though I dropped it the first day and the mirror cracked, which seemed ominous. It played, and still plays, a sweet song that my father identified as "China Nights." I always believed this but wonder now if he made it up on the spot.

No, he didn't. Thanks to Google, I've found MP3 files of the song, and it's the same sweet melody. Further searching reveals its popularity with U.S. Marines during the Korean War and that they often bought highly lacquered music boxes that played the song. Sew girls—the women who worked in barracks, repairing and tailoring uniforms—would sing the tune for their American customers, and they told those customers that "China Nights" was a Japanese soldier's lament for his Chinese sweetheart. A little more research tells me that the music was "written by Shinko Takeoka, and Hamako Watanbe sang the first version. The words were written by Yaso Saijo in 1939 (it is uncertain if Saijo was ever a soldier or ever in China, but the sew girl would probably like to think he was)."

I wish I knew the itinerary of that trip. The camphorwood chest almost certainly came from China but where, exactly? Hong Kong? I know my father was there. In the mid-1990s, he arranged an envelope among the branches of our Christmas tree, and when my mother opened it, she found plane tickets to Southeast Asia, to places my father had gone to on that particular trip and to which he wanted to revisit in her company. Thailand, Hong Kong, Singapore—these were ports of call that still held romance for him. My parents went off in winter to tour those old haunts, sending postcards of beaches and golden temples.

When we returned that day from CFB Esquimalt with the stranger who was our father to our house on Eberts Street, my parents went into their bedroom and we were asked to leave them alone. I imagined my mother twirling for my father in her new suit and then the two of them hugging on the bed. Her Harris Tweed coat was hanging in the front closet, and I went in, closed the door from the inside, and put my arms into its satin-lined sleeves where I could smell my mother's Avon underarm deodorant mingling with the wool. I was inside her coat, inside the embrace she was now sharing with my father. I was my mother, hidden from her children, the collar of the tweed coat rough against my neck.

I still have the carved chest. For years my mother stored all her sweaters in it, and they had the distinctive smell of camphorwood. There was a shallow inner box that sat on a ridge around the top. She kept small containers with various pieces of jewellery on the shelf, and gloves. I keep my sweaters in the chest, and the linen tablecloths that have come from John's mother (embroidered with brilliant flowers by his grandmother in Suffolk), as well as several from the Goodwill on Pembina Highway in Winnipeg, bought while I killed time between readings on a book tour in 2001. I keep my pashminas there too, a kaleidoscope of them, many of them gifts from my children. Everything that comes from the chest carries the smell of my childhood, sharp and arboreal.

And I have the coat, still in good condition, though the cut isn't flattering on me. But a few years ago, when a book of mine was nominated for a prize, I had a brief conversation with another shortlisted author, a man who'd written a book about his father, remembering him through the process of tailoring his father's suit to fit himself. I mentioned my mother's coat, and he said, Why not have it remade? I began to think about what I might do to make it my own.

A thrift shop in the town near where I live always has Harris Tweed jackets on the rack, and I think I know why. I look at people shopping in the grocery store or waiting in doctors' offices, and I listen to them speaking with English accents, or soft Scottish ones. It's an aging population in that town, and when the men die, I imagine their wives asking sons if they want their father's well-kept jacket. Most younger men wouldn't be interested. So the jackets are carefully folded and donated to the thrift shop run by the auxiliary to the hospital. The auxiliary raises funds to buy equipment that helps our small hospital to serve a fairly isolated population. Defibrillators, pulmonary devices, an MRI machine, and I believe the auxiliary even equipped a hospice unit, a birthing suite. I can't bear to leave the nicer of the jackets on their hangers and gladly buy them for $4 a piece. I have in mind a quilt, squares of alternating tweed and velvet, using jackets and dresses from the thrift shop. But then I buy the garments and remember my mother's assertion that she would have her coat for life, and I can't bear to cut into them. "What do you do with the jackets that don't sell," I ask the women working among the old clothes and shelves of mismatched china. I thought that I'd feel better about cutting into one if I thought it was going to the landfill if it didn't sell or else go off to be recycled as rags. "Oh, we make up parcels for homeless people on the Downtown Eastside," one of them tells me cheerfully. So then the thought of cutting up a perfectly wearable garment seems doubly wicked, though I have to say that no homeless person I've encountered in Vancouver—and we often stay in a hotel on the corner of Pender and Homer, walking out on evenings to a play at the Rickshaw or Firehall Theatre, passing the people camped out under the

Army and Navy awnings—has been wearing a Harris Tweed jacket, dapper with jeans or rough corduroys. My husband has at least three of the jackets; each son has several; and dinner guests often leave with a jacket in hand after a discussion about tweed has them confessing to always having liked the look of a tweed sport jacket and a tee shirt underneath. Or a turtleneck. A nostalgia for professors reading from a battered copy of Milton or Byron, or aging gardeners in the west of Ireland, men digging potatoes in a tweed jacket and gumboots, collarless shirt of rough linen or cotton, blue stripes on a creamy ground. I knew some of those gardeners from a time when I lived on a small island off the Irish coast; I sat with them by turf fires, drinking tea so strong that it curdled my throat; and I know that their tweed jackets were never cleaned but carried the odour of bog smoke and chickens, released by rain.

My mother's coat could be made modern and fresh, an elegant jacket to wear with jeans, the surplus from its long length turned into scarves for my sons. There would be scraps leftover, too, to make into something like a token, a heart to fill with lavender to put under my pillow while I sleep.

---

5. No Relation

It wasn't the Grenfell Mission. A simple search on the Internet turns up a document, A Methodist Missionary In Labrador, by one Rev. Arminius Young Missionary to Labrador. (1903-1905). In this account of his missionary work among native people in Labrador, Rev. Young meets Dr. D.T.C. Watson who is a physician with the Nova Scotia Lumber Camp operating in Paradise or thereabouts.

How young he looks, and how certain—of himself, his worthiness, his ability to know what's best for others. He had two children of his own: Helen and Kenneth. Helen never married because her mother refused to let her go; and Ken did marry, and he fathered two daughters. Maybe other children I'm not aware of, but the daughters were the apples of Helen's eye and much praised in the presence of my brothers and me, though we never met these paragons. When we visited, we saw the photographs of Dr. Watson on the table, saw the piano he'd played, its surface damaged in the Halifax explosion. We weren't allowed to touch it, but I imagined the music contained in it. Hymns, no doubt, but maybe old sweet melodies a man might play to children gathered around his feet. One of them not his.

What am I hoping for? This man was no relation, yet I try to imagine my way back to Paradise, or Jamaica, or further back to a Scottish island as steeped in smoke and story as my mother's tweed coat. No relation. My mother had a family—or she didn't. During medical appointments, when I am asked, "Any family history of cancer, heart disease, diabetes?" I can only reply, "I don't know," apart from my parents' medical details. The genealogical listservs are silent when I post queries, the shame of a child born out of wedlock in 1926 vibrating across the decades. Rocking the family boat, on the banks of a buried stream. Shaking up the status quo. All those years when she wondered, hoped, stood stiff-legged with the household dog for a photograph, and in every note, even her sister's obituary, she is foster, not true. My mother deserved so much more than she ever received, grateful for every tiny morsel of affection, too grateful. I remember conversations in the house on Chestnut Street in which my mother's part was less than daughter, slightly more than maid. No one from that house went to her wedding. No one sewed a dress for her, or a trousseau nightdress, with scraps left over to sew into sachets, though in later years, pot-holders arrived in a Christmas box, made with faded cottons, along with an apron pieced together with pieces of ecru linen, decorated with crocheted lace.

---

6. Daylighting

"Daylighting—the practice of restoring a stream that had been routed through a culvert back to its natural state—is becoming a more common stormwater trend..."

Let me bring you back to life, to light. Your parents, your grand-parents, the lines that gave you your black hair, your bright eyes, the spiral of your DNA. Let me trace the route of all those undercurrents, flowing and spiraling, the dark waters idling their way to the sea, find the bed and the riffles of oxygen, the small tributaries that lead away from a source but which might, with effort, allow me to find the spring of your origin.

---

7. Cape Breton Deaths

Every morning, the Deaths arrive in my inbox. Dillons, MacKays, MacNeills, MacLeods. My heart does a little flip when I see a MacDonald or a MacDougall, rises a little into my throat as I scan the obituaries to try to make sense of the dates. Could this be a sister? A nephew? How would I know?

The women all loved to bake and never let a person leave the house without a meal. They knit, they grew flowers, they played a mean game of bingo. The men were good providers. Had big smiles. No visitation by request. Or: Everyone's invited to the wake. My mother loved the music of her home province and would have clapped at these wakes, I'm sure, the fiddles and accordians her connection to "home" each week as she watched *Don Messer's Jubilee* while ironing. More than once, I saw her cry as the final song played:

Every tear will be a memory,

So wait and pray each night for me,

Till we meet again.

Reading the obituaries of MacDougalls and MacDonalds, I keep wondering if they are related. A man who worked in the mines, who loved poker, who doted on his grandchildren. A woman who went into nursing, who worked at the Dominion Store, the fish plant, who will be remembered for her humour. "She was an avid bowler who lived a simple life devoted to her family and friends. To Sis there was nothing like a good game of scat and a cup of tea." A grandmother, mother, aunt, daughter, who loved cribbage and flea markets—this one I read again and again, searching for something resembling a deeper clue, a hidden message. The hidden tribes of Cape Breton gathered by the buried streams of her birthplace, holding a seat around the fire for her, a cup of tea kept warm on the stove, the cards laid out for the grand game.

---

8. Rings

"It is also the case that some parents used the token as a means of expressing their feelings on parting."

I have their rings—his was sealed into a small envelope with "Tony's Ring, Centre 3" written on it, indicating the ward where he died. When we were clearing out their apartment after my mother's death, one of my brothers wondered about our father's ring. We assumed it had been lost, and I don't recall putting this small envelope into the boxes of papers I brought home with me. But everything was chaotic, muzzy with grief. And so now here's his ring, many layers of cellophane tape wrapped around it so he could keep it on his finger towards the end of his life when he had faded away to almost nothing, and then nothing. My mother's rings lay in the tackle box where she kept her jewellery. One of them, her engagement ring fused to her wedding band, rests in a small box, like the kind dentists send you home with when they've removed your wisdom teeth and rinsed them in the sink. Both bands had worn so thin over the years that she must've taken them to a jeweller to have the delicate circles welded together so she wouldn't lose one without the other. Her rings were sacred to her. When I married, I had a ring, a silver ring with a moonstone, but I seldom wore it, and then I sat on it—in the back pocket of my jeans—and flattened it. (It could have been a portent but it wasn't. We celebrate our 37th wedding anniversary in a month.) My mother was horrified that I didn't wear a wedding ring. How will anyone know you're a married woman, she asked. And I said, Well, I know I am, so that's all that matters. But to her, the ring demonstrated to the world that you had found true love and made it work. That you were respectable, beyond reproach. You could hold out your left hand, and the world knew you weren't on the shelf, an unclaimed treasure, a spinster. A foundling waiting for her true mother to claim her.

Mum and Dad, their wedding day, August 4, 1950

So I have their wedding bands, my mum's engagement ring, and also her previous engagement ring—from Bernard, the man who loved her before my father came along, and whose photographs fill a small black album, a dashing guy who signed them to her All my love, Darl. Imagine abbreviating Darling! He must have been breezy and confident, which my father wasn't. Yet she cancelled their engagement, and he insisted she keep the ring, which is curiously like her next engagement ring, a simple diamond in a solitaire setting. Did she choose them herself, or was she presented with a ring in a small velvet box, and was the simple solitaire what she inspired in those who loved her?

And then her family ring. For a while, those were big, featured in every Christmas flyer, and occupying at least a full page in the Sears catalogue—rings with the birthstones of every member of the family. My father bought her a family ring as a Christmas gift but its stones were surprisingly dull and generic. The ring is not beautiful, and I can't imagine ever wearing it, but I also can't imagine giving it away, the adjacency of birthstones a charm to keep the family together.

---

9. "And I will give unto thee the keys of the kingdom of heaven..." MATTHEW 16:19

I knew you were awake when I came in late. Before I left home, I slept in the basement, but that door was always triple-locked, with a two-by-four set into brackets bolted to the walls beside the door, because you and Dad were certain someone wanted to break in and take what you had. Your 1950s hi-fi, your aging television cabinet on thin legs with the wobbly picture, the fridge that wheezed in one corner of the kitchen. The jewellery you kept in a tackle box under the bed, everything from a few nice jade pieces from the Far East to cheap kiwi earrings you wore with any outfit. Maybe the bottle (any liquor was called "the bottle"; there was seldom more than one, though occasionally two, gin and Scotch) tucked into the back of your closet, which always smelled of socks, a legacy of Dad's naval years when athlete's foot ran rampant through the showers of the *Ottawa*, the *Restigouche*, the *Stettler*. Most windows of the house had Plexiglas screwed onto the frames, as an extra level of security. Keys were given with admonishments not to lose them, to guard them carefully.

So I had to come in the front door, my key loud and awkward in the lock, which could have used some graphite to ease its workings, and anyway the sweep of my car's headlights would have shone right into your bedroom when I pulled into the driveway. I never wanted to talk. My social life in those years was desolate. A handful of friends, almost no men who'd look at me twice, let alone ask me out, apart from aging and alcoholic painters and poets who dreamed of youth in my arms. I'd go down the dark stairs to my room that looked out into the carport and try to dream my way to happiness. I knew you disapproved of me, and if you knew how I'd spent my evenings, you'd disapprove even more.

You hoped for something other than this. You hoped I'd find a nice boy with a good job who'd take me to Birks to buy a diamond ring and set the date. Maybe by then, I'd be a teller at the bank in the shopping centre near where we lived and could flash my ring to all the other tellers. I was difficult to explain to your friends. Well, she writes. Poetry. We're not quite sure what she'll do next. She's borrowed her dad's truck to go camping alone to Englishman River. We tried to say it was too cold in November, but that girl never listens to us.

I wish I'd been better at making you happy. Sometimes we went shopping together, and you tried to get me to look at clothing in bright colours. You couldn't understand why I always bought black. It matched my mood, and anyway, I'd heard it was slimming. If I was thin, maybe one of the beautiful young men I dreamed about would look my way and see me for the first time, forgetting the chubby girl who sometimes drank too much and talked about the spirit. Or god help us, the soul. Or the weight of history in a line of Seamus Heaney. Who, when prompted, could recite "When in disgrace with fortune and men's eyes" by heart, a party trick not much appreciated.

I knew you were asleep when we came in just a few days before you died, our key quiet in the lock. Your arms were blue with cold, and thin, so thin. I tucked them under your comforter and made sure your oxygen tank was functioning. I put roses on the table and sat in the dark and cried. *For thy sweet love remembered such wealth brings...* It wasn't always sweet, not always welcome, but it was wealth, and then I was bereft.

---

10. The Nobodies

Some days, she was Mrs. Nobody. How airily she'd say that, and of course it meant nothing to me. I never parsed the sentence, her too-bright smile. And some days, the Girl From Sooke, also said airily, a person who washed dishes in the one sink, putting them in the blue plastic rack, then dried them one by one with a linen cloth printed with lobsters from Peggy's Cove or wild roses from Alberta. The Girl From Sooke, who lugged the laundry bag downstairs to sort and wash our clothing. And who polished the wood furniture with Pledge.

Mrs. Nobody sat at the kitchen table with her cup of instant coffee and an Export-A, wondering why the house never stayed clean, why it was so hard to make ends meet, why the dollars never stretched far enough. Her *Redbook* magazine helped her prioritize—put so many dollars aside for food, shop in bulk, can this marriage be saved?

Mrs. Nobody, daughter of ... Well, who were those Nobodies that made this woman with her black hair and sharp eyes, her perforated eardrum from a childhood illness no one diagnosed, her shapely legs (which I most certainly didn't inherit), and her narrow shoulders (again ...). The Nobodies, of Nova Scotia, one of them the brother of a doctor, maybe the doctor of the telegram ("Give her to Dr. Watson's widow. She'll be discreet.").

---

11. Their Chair

I looked out just now and saw their chair by my bedroom window. Part of the patio set they won from a radio show, made of dark green webbing with textured cushions. We brought it back from their apartment after my mother died, and it's the one my husband sits on when we have coffee on the upper deck, the deck that surrounds three sides of our second-storey bedroom and his study and our bathroom. Three sides of weather and treetops and the mountain. Usually the chair stays over by John's study, but after lunch I carried it to the small area in front of the sunroom door where the dog rose climbs around the bedroom window, along with trumpet vine, wisteria, and deep pink honeysuckle. The sun travels lower on its trajectory from east to west, from Hallowell to beyond Texada each September day, filtered more densely through Douglas firs than even a week ago; in high summer it passes directly overhead, clear of trees, from its rising at 8 until its setting at least twelve hours later. I sat in the chair for half an hour, in-between watering and making tomato jam, rereading *Portrait of a Lady*. And I then returned to work, because sitting felt too much like indolence. The chair smells of them. Passing it now, after its few hours in late sunlight, I can smell my parents as though they're both there, taking the warmth of an afternoon, talking quietly, not noticing me in my old skirt and tank top, hair wrestled into a knot to keep it from my face as I reach into the tomato vines for more fruit for my jam. I never knew I would miss them as much I do now, smelling them in the coarse green cushions, my book abandoned across the seat. There was so much I never told them. They didn't want to know about books (Henry James?) or lofty thoughts or travel plans for Europe. They hoped I'd share ideas for stretching a dollar, ways to shop thriftily, to use up odds and ends from the fridge. Varicose veins and sore teeth. Stomach acid or the wisdom of generic vitamins or difficulties with the bowels. They wanted me to prove I was theirs, that I'd paid attention to their lessons, their advice, that no one else meant more to me than them. It's taken me so many years to learn that there is some truth to this. I look at my hands and see hers. My slow metabolism and sluggish blood pressure come from him.

Later, I look out again, the back of the chair impossibly sad. His head touched the green vinyl strapping. Hers, too. On the shabby deck of the house on Mann Avenue, they sat in their chairs—this one, and a kitchen chair brought outside through the sliding doors that they locked after each use, bolting down the extra Plexiglas panel at night against all those who lurked, wanting to break in to steal their hi-fi, their clock radio—waiting for the seagull who came some days for old bread. Willie, they called it. Also a neighbour's cat. The heavy foot of the mailman as he trudged up their stairs.

———

12. What lasts: A meditation on My Sin

When my mother died two years ago, I carried home the camphorwood chest that my father brought back for her from southeast Asia in 1962. I was in grade one then, and it was the first time I remember my father being away from us for any length of time—three months, or maybe four—a long time for my mother to be alone with four children, no car, and no contact with her husband, before inexpensive phone calls. He wrote letters. He sent postcards to my brothers and me. And he brought home the chest filled with gifts.

One of the gifts he brought my mother was a bottle of My Sin Eau de Lanvin. She kept it in the trunk and wore it for extra special occasions. She'd dab a little behind her ears, on her wrists, and on a little cotton pad that she'd tuck into her bra. Once when I was about 15, I went surreptitiously into her bedroom, opened the trunk, and soaked a tiny cotton pad with My Sin that I tucked into my own bra before going to a school sock hop. I felt so sophisticated. I imagined boys swooning at my feet. But not a single one asked me to dance.

I have the bottle of My Sin on my desk as I write. It's still in its original box, though the top of the box is missing. Paris France is printed across the bottom of the box and a circle with 85°. On the back of the box: 225 GR.Env.Par Flacon. The bottle is three-quarters full of an amber liquid, the Sin itself.

I imagine my father bought it at a duty-free store. Did he sniff a number of perfumes and did this one speak to him of my mother, whom he loved, and the woman she was to him, elegant and alluring? Before they had children, I think they went to a few fancy dances, but that ended with our births. There wasn't often enough money for babysitters or new dresses. But very occasionally she'd dab on her perfume and put on her muskrat coat, check the seams of her nylons to make sure they were straight, and out they'd go, into the night, her scent drifting back to me. As it does now, sniffing the cap of the bottle.

I learn more about My Sin through some of the many sites on the Internet devoted to the history of perfume: "Lanvin My Sin (Mon Peche) was created back in 1924 by a mysterious Russian lady called "Madame Zed," who worked on several more Lanvin fragrances. This feminine, provocative and dangerously seductive fragrant composition begins with aldehydes, bergamot, lemon, clary sage and neroli. The middle notes are: ylang-ylang, jasmine, rose, clove, orris, lily-of-the-valley, narcissus and lilac. The base is oriental—woody with vetiver, vanilla, musk, woody notes, tolu balm, styrax and civet. The perfume was discontinued in 1988, but it is still available on line."

What do I do with a bottle of fifty-year-old perfume? I am 57 myself. It's not something I'd wear. I discovered Chanel 19 in 1972 and have never found any reason to change. I don't even know if this bottle is still viable. Does perfume turn to vinegar, as an opened bottle wine will if not used within a reasonable time? When I sniff the bottle cap, I say that I smell my mother but how can that be? She wore perfume so seldom—one-quarter of a bottle over forty-eight years. Maybe she knew she would never have another bottle of French perfume, maybe she wanted to ration it to keep the memory of my father's return fresh. What I am smelling is the way I would like to remember her, in a rustling cocktail dress one or two evenings only, her feet wiggling into pretty shoes, checking her seams in the bedroom mirror, her eyes bright with anticipation of dancing! Not the old disappointments, a daughter who didn't visit often enough, the house sold, her husband dead, the days growing shorter and shorter as the year approached the longest night, the bottle of French perfume forgotten in the camphorwood chest, among the gloves and her one cashmere sweater, an old silk square from Zanzibar folded neatly on the bottom.

13.

Under Cape Breton's rocky soil, under the parks in Halifax with their views of the sea, the sound of gulls, of commerce, of pianos and fiddles from open windows, under the earth the buried creeks hide their secrets. And you can hear something, a murmuring, a rill of original water, of origins, of fish in their lost habitats, eels, amphibians entering their dark waters, and in memory, birds at the vanished banks, their beaks poised, and secrets, secrets, my mother's buried history in the damp ground where water longs for the sky.

———————

14. Burning her recipes

A small black scribbler, the cover textured like cheap crocodile skin. Her handwriting on every page, and some splatters of oil, grease, pancake batter. Yet I don't remember her using the scribbler when she cooked. Her repertoire was neither large nor fancy. Macaroni and cheese one night a week, after 1966 spaghetti made with meat and tinned mushrooms because a neighbour had recommended this exotic dish as a perfect way to stretch a pound of ground beef; fried chicken, which I loved; a mixture of ground beef, canned peas, and onions over riced potatoes; a roast on Sunday, put into the oven before church so that when we arrived home, the house smelled of beef or lamb stuck with slivers of garlic; goulash, which was much like the spaghetti sauce, flavoured with paprika instead of oregano. Once she and my father bought a Kraft pizza kit, and she followed the instructions on the box. My younger brother and I cried at the smell of the powdered cheese. Like throw-up, we sobbed, and she told us we didn't have to eat any of it but there would be nothing else.

In an envelope, a mess of newspaper clippings, all recipes. A fruitcake I don't believe she ever made. A casserole of chicken and dried apricots that I think she did try once when we were older. My father professed to hate chicken, so it didn't appear often. Just the fried version when he was away at sea. A handful of cards from a magazine with tips for the busy housewife: bright images of pork chops glazed with apricot jam or syrup; meatballs in tomatoes and wine, heaped over noodles; a cake made from a boxed mix, but made richer with pudding and fresh eggs, then topped with Dream Whip and crushed pineapple.

All up in smoke, along with photographs of her foster sister in a nursing uniform, circa 1935, some Sunday School classes, sailors (my father among them) in costume under palm trees in some exotic port, and people I couldn't recognize, so didn't want to keep on the off chance (a phrase she used often, a justification for keeping so much, in such disorder). You could recycle some of this, my husband said, but I wanted it gone, out of the house, and the day was warm, some cloud, no rain: a good day to burn the past. My heart did not have room for faces too far away to touch. For clipped instructions to feed a family of six. For finger-prints in ancient batter on the cover of a scribbler.

Now that the fire has died down, I can smell the smoke in my hair, the old recipes and their attractions reduced to fly ash, the photographs shrivelled one smiling group at a time. That night I will wake from a dream of my mother, walking alone, her recipes burned, no one to make dinner for any longer, and the heavy scent of smoke upon my pillow.

15. No note tucked into a nutshell or locket or thimble

"When the tokens now on display were removed from the billets in 1850s and 60's, the links with the children with whom they were left were broken."

I visited the Foundling Museum again in 2015, walking over with my husband from our Bloomsbury flat to attend a concert on the occasion of Handel's birthday. The ticket price to the concert included the opportunity to look at the cases of tokens, and I was glad to walk their lengths in a quiet room, seeing again the hairpin ("left with a boy who became Humphrey Joyce and who survived to apprentice as a needle maker in London"), the gold buckle left for a boy who died just ten weeks before his mother came to claim him, the thimble, the bone fish, and the tiny brown hazelnut.

Having sorted my mother's boxes, I am no longer sure that any token exists to link her to a mother. A foxed telegram? A dead end, or at least a path into a world of the MacDonalds and MacDougalls of Cape Breton where the bowlers and tea-makers have not opened their circle to admit a girl born to one of them in 1926 but still abandoned. No birth certificate, no note tucked into a nutshell or locket or thimble. Nothing hidden in plain sight. Not that I could see.

Sitting in the beautiful Picture Gallery at the Foundling Museum, listening to Charmian Bedford sing lute songs

composed by Handel and Purcell with musicians from the
Academy of Ancient Music, I
lost myself in the sad tangle
of connections.[2] I wanted the
music to be coded, wanted to
believe that messages might
come in its ancient verses,
as I wished for a message to
emerge from that telegram,
an invisible milk or lemon
juice exposed to heat: "Tell
me, some pitying angel tell,

Mum in wicker chair, circa 1931

quickly say, Where does my soul's sweet darling stay?" But I
never thought of her as a sweet darling, mostly because I don't
think she ever thought of herself that way. She was nobody's
beloved child, though she had solid breakfasts, warm clothing,
a clean bed at night.

> Why, fairest object of my love,
> Why dost thou from my longing eyes remove?
> Was it a waking dream that did foretell
> Thy wondrous birth?

No one regarded her birth as wondrous, though perhaps in
later life her birth mother thought of her with something like

curiosity, something like love, regretting that she gave her up or at least did not seek her out to explain or simply to know her after years of absence. A girl, a young woman who could twirl back into her life in a powder-blue suit, a tweed coat. An old woman alone in a chair by a window in Victoria. Seek her out, know her, grieve for her. As I now grieve for the absence of her company, which I confess often made me irritable, impatient. Her complicated stories, straying far from the point. Her bright eyes never missing a thing, though I did not celebrate such careful attention but bristled to its scrutiny.

I have written elsewhere of a paperweight she remembered from her childhood, something she was responsible for dusting weekly in her foster mother's living room, along with a piano damaged in the 1917 Halifax explosion, and a desk containing the journal of that earnest young medical missionary whose presence shadowed that house. The paperweight sat on the desk, and it inspired dreams in my mother; after her foster sister died, it was the one thing she wanted as a memento. But the neighbour who'd inherited the contents of the house said her husband had "taken a shine" to it (a shine made possible by the years of my mother's dusting) and wouldn't relinquish it. In it, she saw possibilities of another life: I wrote, in an earlier book, "Whenever my mother dusted that room, she would linger at the desk, looking at the ornament which she knew was precious to her mother (sic). I heard her say that the paperweight was linked in her mind with a story she'd read about some children on a mission to Norway where they had to deliver a paperweight, though she couldn't remember to whom or to what."

So now I have tokens, left in the event she should return to claim me, in all my imperfections—a child who burned recipes, who resisted sitting on her bed to share details of her life, a life I thought she'd disapprove of, but maybe I would have been surprised. Was I the fairest object of her love all those years when I felt myself homely, lonely, my face too dark, my legs too thick? Did her longing eyes seek me? Was my own birth wondrous to her. I doubt it. She was alone with two young sons, my father at sea, as he would be for so much of my childhood. I've searched for her mother, who never returned, who never claimed her in word or deed, but maybe I should have concentrated more on her. Her true heart, her own plain virtue.

At the Foundling Museum, a spyglass, a hairpin, the handle of a penknife. Padlocks, a tiny black hand pierced with a hole for a ribbon, a handful of coins, pierced, notched, worn thin by thumbs stroking, stroking, stored in the archives. I have My Sin, a tweed coat, a memory of Mrs. Nobody on her chair in the kitchen. I have a hole on my sleeve the shape of a heart but no scrap to match it with and the sound of a creek running underground on its way to the sea, with everything of my mother in it, and nothing. I have every regret for the way her life began, and ended, a motherless child, so far, so far from her home, no one looking for her in the listservs, among the dry records of Vital Statistics, no one, no one but me, my face against the glass case of all those unclaimed tokens, those stories begun perhaps in love and ending in sorrow.

NOTES

1. The epigraphs from *An Introduction to the Tokens at the Foundling Museum*, by Janette Bright and Gillian Clark. London, England: The Foundling Museum, 2011, *Threads of Feeling: The London Foundling Hospital's Textile Tokens, 1740-1770*, by John Styles London, England: The Foundling Museum, 2010, and "Daylighting Streams," by Carol Brzozowski, in *Stormwater Magazine*, September 1, 2010.

2. Passages of Henry Purcell's "The Blessed Virgin's Expostulation" (1693) are quoted in the last section of the essay. I heard Charmian Bedford sing this, accompanied by members of the Academy of Ancient Music Orchestra, at Handel's Birthday Concert at the Foundling Museum in February, 2015.

My father on a trike in Drumheller, circa 1929

· · · · · · · · · ·

WEST OF THE 4TH MERIDIAN
A Libretto for Migrating Voices[1]

═══════════

"Scientists have sought for centuries to explain how animals, particularly migratory species, find their way with awesome precision across the globe... Dragonflies and monarch butterflies follow routes so long that they die along the way; their great-grandchildren complete the journey." M.R. O'CONNOR[2]

───────────────

1.

I am holding the family song, a composition almost erased. I carry the long passages into the next bar where there's a rest. A little run of sixteenth notes, semiquavers, dry music in the cemetery we found driving east of Drumheller, an obbligato of magpies. Back and forth, the song goes, though my part is uncertain.[3]

───────────────

2.

I heard stories in childhood, or think I heard them, and they've formed the matrix of my family, its origins. Was it my father

My grandmother Anna with her first husband Joseph Yopek, circa 1900

who told the stories? Or did I hear them in Edmonton backyards where my aunts and uncles talked long into the night while we slept in a tent pitched under a poplar? Phrases, half-remembered songs, names, the prayers my grandmother Anna Kishkan, nee Yopek, nee Klusova, muttered as she sat in a rocking chair nearby and ran her fingers up and down a chain of beads my father identified as a rosary. The names, the voices, singing me to sleep under worn blue canvas, as the recitatives, repeated phrases, little trills like birdsong, a deep basso profondo like an oktavist in a Slavic choir, echoed the languages I heard only among our extended family in Alberta, apart from once or twice when my father let me join him at Mass. A migration from one continent to another, from a language to another, a landscape of beech trees and foothills of a mountain range stretching across Europe, another matrix once described by Ptolemy in his lost atlas *Geographia* as Carpates. There were mountains in that range known by both grandparents, in separate countries, in different languages. And they arrived in the badlands of Alberta, dry, dry, with rivers tarnished in sunlight, frozen in winter. I never questioned the truth of the stories, though the detail was scanty.

In April 2016, we traveled to Edmonton to visit our son Brendan, his wife Cristen, and their little daughter Kelly. We also planned to drive down to Drumheller to try to find traces of my father's family who settled there in 1912-1913. My grandmother Anna's first husband, Joseph Yopek, came to North America before her own arrival in 1913. I'd understood that Joseph took out a homestead grant in the Drumheller area where he'd also found work in a coal mine. (His brother Paul was already in Drumheller.) The family story had him preparing a home for his wife and their five children, brought from Anna's birth village of Horni Lomna in what's now the Czech Republic.

I have a handful of photographs from the 1920s, taken on what I suspected was the land where Anna and Joseph settled and that my grandmother must have inherited after Joseph's death during the 1918 Spanish flu epidemic. A funeral for Julia, the first child born to my grandmother and her second husband, my grandfather John Kishkan, in 1921 and dead of diphtheria in 1923. My father, Anthony Kishkan, known to his familiars as Tony, on a small trike in a rough yard with a dog. Another of my father in a little car with some wash tubs stacked behind him and bleak hills beyond those. I wondered if this was the land. Dry, dry, and a river nearby.

My father talked of a farm, chickens, a cow whose milk my grandmother turned into butter to sell during the hard days of the early 1930s. She did the farm work while my grandfather worked in one of Drumheller's coal mines.

When I searched the Alberta Homestead Records for 1870-1930, I found Joseph Yopek and a description of the quarter section in his name: Section 11 Township 29 Range 20 west of the 4th Meridian. At the Provincial Archives, I wanted to obtain a copy of the patent or title in order to determine the location of the land, so we could look for it in Drumheller.

I gave the archivist the file number, and she set up the microfilm for me. I expected to find several pages—the patent letter, perhaps a map. Instead, I found nearly 400 pages of documents, beginning in 1909, ending in 1928, which detailed a surprising saga. The land I thought Joseph homesteaded was instead a squatters' settlement. I couldn't determine how many lived there. I read through petitions that asked for permission to buy small plots.

(We the undersigned petitioners, being the present occupants... Humbly pray.)[4]

I found no family names on that petition, but my grandmother's first husband and his brother spoke almost no English. They were functionally illiterate, my father told me. The whole family, he said. They couldn't read or write. His own father could make an X and later my father taught him to write his name in round uncertain letters. I read directives from the

Department of the Interior warning squatters not to assume that they will be compensated for structures (their homes) built without permission. In these files, I discover that Joseph and Paul did receive a letter, though their names are misspelled. Reading further, I discover they are listed among the names of 22 squatters living on the SE quarter of Section 10 and another list, 41 names, associated with the SW quarter of Section 11.

Promises and directives went back and forth between Alberta and Winnipeg and Ottawa. There were partial maps and suggestions that blueprints for subdivisions were being prepared. 1915, 1916, 1917; the lots were finally posted for sale on November 12, 1917. Prices ranged from $25 to $400. The Canadian Northern Town Properties, a real-estate subsidiary of the Canadian Northern Railway, acquired many of the lots.[5]

I read what I could and printed the pages I thought might be important, but I never made it to the end of the file. Later, my son in Ottawa copied the entire file for me and sent it as a pdf. I sit at my desk now and try to make sense of the story. Joseph and Paul's scrawl moves me to tears—rounded and shaky like my grandfather's later signature. Paul either disappears or else finds another place to live. I find the name of a man who is almost certainly my grandmother's brother, Joseph Klus, though my father never knew she had a brother.[6] My father never mentioned the details of Joseph Yopek's death. I am sure of this.

"Did your parents have brothers and sisters?"

"I never asked."

Never asked. None was ever mentioned.

"I walked those hills as a boy but never found anything worth keeping."

Never knew or never told that the family home was a shack in a former squatters' settlement. It begins to sound like an aria. Never never never never. Think of this as the refrain. Never. A fermata over it so you can hold it as long as you can. As long as you need to. Never. Ever.

No Yopek nor Klus on the list of successful purchasers. I expected the record of a home, or maybe just the memory of it, its contours on a map, a familiar view. Those hills beyond the washtubs... I think of the frigid winters and the sound of coyotes, children fishing off the cold iron bridge, the bins of potatoes dwindling. Chickens long since consigned to the soup pot. Sound a low note for hunger, the second F below middle C.

In the Depression, remembered my father, my mother sold any kind of food she could. Noodles. Butter. Those chickens. Homemade cheese from the cow, kept until they could no longer afford winter feed and no children remained at home to drink the milk.

What was my grandmother praying for while her grown children talked and sang in that Edmonton yard, many of them already grandparents themselves? While those aunts and uncles made us stand back to back with other children, comparing our height, the shape of our faces, looking for common cheekbones and noses? My father said the rosary was divided into mysteries—sorrowful, joyful, and glorious—and that you held the beads and remembered. My grandmother's hands were bent and knotted, and the beads clicked as her knuckles clicked. She held a crucifix to her heart.

Tufted fleabane, prairie crocus, early yellow locoweed, showy milkweed. Old World swallowtails, sulphurs and marbles, anglewings and checkerspots. Turn, turn, bend the song to the roadside plants, the hosts, long syllables, short ones, free verse composed of craneflies, dragonflies, bluebottles, broad-bodied leaf beetles, greasewood and cocklebur, the miraculous monarchs hovering.

I am holding the family song, a composition almost erased. I carry the long passages into the next bar where there's a rest. A little run of sixteenth notes, semiquavers, dry music in the cemetery we found driving east of the town, an obbligato of magpies. The graves we found on old maps—Joseph Yopek; two babies, Myrtle and Julia; and, Joseph Klus, who came by boat a few months after his sister, my grandmother, Anna. Never heard from again. In Horni Lomna, where Anna was born on a little farm on a slope of the Beskydy Mountains, which stretched east to the Carpathians where my grandfather grew up in Ivankivtsi, there is a record of two sisters but not Joseph. I have Anna's rosary but not her faith. I have her strong legs and her ability to grow vegetables. Back and forth, the song goes, though my part is uncertain.

———————

3. Both Joseph Yopek and Joseph Klus were coal miners...

"They were told that they might remain on the land as winter was coming on."[7]

———————

4. What is your name in full, age, occupation and Post Office address?[8]

"I was 40 years old when I died. I was a coal miner. I had a Post Office address. Who would I have written to, but Anna? Who would have written to me? Nothing survives."

I never knew you, Joseph Yopek. But you were my grandmother's first love, a man from a village over the border in Poland. In the family song, you were hardworking, industrious, father to nine children, eight of whom lived. You came to North America alone, leaving my grandmother behind with five children. You were in West Virginia in 1911, but in January 1911, your son Frank was born in Horni Lomna, my grandmother's village in Moravia. Did you stay for his birth? Did you hold him by a fire of coal and black spruce and tell him he would cross water with his mother?

I know almost nothing about you.

But had you written a letter, would you have said that you'd landed on your feet west of the 4th Meridian, that you were Właściciel ziemi or obszarnik (the first a laird, you might have wanted them to admire what you'd accomplished; the second a simple owner of land)? Did anyone explain to you the difference between occupying and owning?

---

5. Is there any possibility that the small fraction of Section 11, North of the townsite and along the South bank of the Red Deer River here, will be sold this spring?[9]

"Squatters are building all over this township. These lands are badly needed for settlers who work in the mines. The buildings being erected are small shack like structures, but if the land could be bought the buildings would be more substantial."

---

6. Who else wanted your land? (sing slowly as a round):

"... will... will you be good enough to take this matter up with Ottawa again and press them for a reply. The squatter situation has developed much to our disadvantage, and would like to know what steps the Government is going to take to relieve it."[10]

7. I'm remembering everything backward.

I'm reading *Death by Water* by Kenzaburo Oe right now, a strangely compelling novel.[11] It took some getting used to—the prose is anything but brisk and lively—but the main character, a writer in his seventies, is reliving the early years of his life. There's a poem in the novel that functions as a sort of mantra—the first two lines were written by the character's mother, and then the short poem is finished by him. He realizes that it tells the story of that relationship, and it also sums up so much of what he is thinking about at the late stages of his career and his life.

> You didn't get Kogii ready to go up into the forest
> And like the river current, you won't return home.
> In Tokyo during the dry season
> I'm remembering everything backward,
> From old age to earliest childhood.

It's a strange experience to be pursuing the sad origins of my father's family at the same time that my immediate family is growing and flourishing. In Edmonton, on the same grey-scale film as these old photographs, oddly enough, I viewed the ultrasound of Cristen and Brendan's baby, due in September. I saw the baby's hand, the baby's face. And last year, in late February, as John and I visited Amsterdam to attend a wedding, a call came to our hotel from our older son Forrest and his

wife Manon to tell us that they
were expecting their first baby.
Moments later, an ultrasound
of beautiful Arthur arrived
on my small Samsung tablet. I
hold all of these in my mind
and my heart's archive, these
grey approximations of the lives
I cherish, even the ones so far
away in time, that I will never
know exactly where the boy
who rode that little car lived, or

My father and car in
Drumheller, circa 1929

where the family gathered in front of a weathered house dis-
persed to after the funeral. And did that boy's grandparents, my
great-grandparents, back in the small house in the valley below
the Mionsi forest, ever see a photograph of him? Ever learn his
name? They never saw their daughter Anna again.

Last night, after dinner, we walked a route that brings us
home through the woods. I was thinking and remembering so
many things on the walk. How the maples are blooming earlier
than usual, and what was that song I could hear (a warbler?), and
how many times did my young daughter Angelica and I collect
tadpoles in the pond we pass, and wasn't that where we saw
monarchs feeding on milkweed, and the last time my father did
this route with us, he could barely walk back. I keep thinking
that if I just pay attention, it will all become clear to me, the old

house, how close it was to the Red Deer River, who slept where within its small dimensions, and how to find my own way to it, dreaming or awake. The place on the bridge where my father fished, his line taut in the current, his eyes green as the water. Dragonflies stung the surface of the river, wings like nets.

---

8. When did you build your house thereon? And when did you begin actual residence thereon?

"I lived with my brother Paul for a time and then had a house ready for Anna and our five children when they arrived in spring of 1913."

---

9. What is the size of your house, of what material, and what is its present value?

"I have a plain house, 20 feet by 25 feet, with a caravan roof (I was told to make the roof this way by a man who knew such things), and I have a garden 80 feet by 125."

---

10. What residence have you performed on it?

"We lived as a family of 11."

In the list of structures on the SE quarter of Section 10 Township 29 Range 20 West of the 4th Meridian, some houses have tent roofs. Others are shacks built of bark. Old wood. Tar paper. Holes in the ground with sod for roofs. A dugout in the riverbank (my grandmother's brother). I try to imagine these dwellings, particularly the house where my grandmother lived. How could 11 people sleep in such a small space? How could the children have done their homework and practiced their English? How could they manage their laundry, especially with several children in diapers at any one time, how did others sleep when someone was ill, how were clothing and mattresses and quilts sewed and mended, and how much light did they have during the long cold winters. (In a bar of music, you could pause, you could rest, but in a household of 11 people, the tempo would be busy, the notes quartered, then quartered again.)

---

11. Anna, saying a Hail Mary with the rosary she brought from Europe:

I hear the beads clicking as I heard them then, as she sat in her rocking chair, chanting, in Latin? In English? I don't remember. She always spoke with such a heavy accent—during her years married to Joseph Yopek, who spoke Polish, or else Cieszyn Silesian, the Czech-Polish dialect she knew from her own village; during her years married to my grandfather, who spoke

Ukrainian, her children all talking and singing in English, the language she went to my father's elementary school in Drumheller to learn, her large soft body in a small child's chair. At home, in the shack on the Red Deer River, the family kneeling for the evening prayer, their rosary beads softly ticking like a metronome. They had so little yet were grateful.

Ave Maria, gratia plena:

Hail, Mary, full of grace;

Dominus tecum,

The Lord is with thee;

Benedicta tu in mulieribus,

blessed art thou among women,

et benedictus fructus ventris tui, Iesus.

and blessed is the fruit of thy womb, Jesus.

Sancta Maria, Mater Dei, ora pro nobis peccatoribus,

Holy Mary, Mother of God, pray for us sinners,

nunc et in hora mortis nostrae. Amen.

now and at the hour of our death. Amen.

---

12. Is there water?

"In regard to the sanitary conditions on the land occupied by the squatters, I find very similar conditions existing as you will find in many of the small Western villages where there is no water works system.

In regard to having the squatters vacate the land I wish to say that the majority of them are poor men who are unable to purchase lots in the townsite."[12]

I cannot imagine the barren shelves, the tin cup for water shared around, empty pockets in mended overalls. Coyotes on the ash heap, the bitter stalks of yarrow.

---

13. What extent of fencing have you made on your homestead, and what is the present value thereof?

"Scraps of trees pulled from the river after snow melt. I sent my boys down with ropes. I begged them to take care."

---

14. What other buildings have you erected on your homestead? What other improvements have you made thereon, and what is the cash value of the same?

"Any persons squatting on this land do so entirely at their own risk and are liable to lose any improvements they have made thereon as they would not be entitled to compensation in the event of the land being offered for sale."[13]

"What is a chicken shed worth? What is a roof on four poles worth, apart from the shelter it gives a cow? My house they say is worth $150. But who would pay this for a house on land no one can own?"

---

15. Tell, tongue, the mystery of the glorious Body (to be chanted in the plainsong version of the Pange Lingua).

"Understand another sale of school lands is to be held
in Alberta soon   the town   council of   Drumheller urge sale

of southwest quarter section eleven twenty nine twenty as
present conditions are a menace   to towns   health
and finances

as we are practically debarred fromgetting any revenue from
people on this land   you had already agreed to the

sale of this parcel of land and without warning you
withdrew it conditions demand quick action and the
town expects

a favorable reply at an early date"[14]

16. Are there any indications of minerals or quarries on your homestead? If so, state the nature of same and whether the land is more valuable for agricultural than any other purpose.

"Perhaps coal. But what is more valuable than a place to live and raise your children, to have a roof over your heads, soil for the potatoes and cabbages, a river for its water brought home in buckets? We had a latrine dug into muddy ground. We had oil lamps and tallow candles Anna made from beef fat she begged from the butcher in exchange for holubky or a bowl of pierogi stuffed with sweet plums, a little bowl of smetana alongside."

17. How many horned cattle, horses, sheep, and pigs, of which you are owner have you had on your homestead each year since date of obtaining entry? Give the number for each year.

"There was a cow. Anna made butter to sell. There were chickens, as many as she could keep. Anna left some eggs with the hens to hatch and grow. She went out with old bread and the ends of cabbages, potato peelings, and the chickens gathered around her feet. When our family went hungry, so did the chickens. They ate grasshoppers in summer and caterpillars that fell from the willows. Anna made noodles to sell, and cottage cheese, fresh bryndza; miners I knew lined up at the door for the yellow kluski she dried on the handle of her broom."

18. How does a neighbour, Mrs. Bond, describe her life?

"World War 1 started in August, 1914, and on October 2 my second baby was born. We called him Tom. He was only a few weeks old when my husband was laid off, so we had to leave our home because it was a Company house. My husband got lumber and built a small place on the School Section nearer the town, similar to those being built by a number of other people. The houses were longer one way than the other, and could be converted into two rooms. They had a caravan roof, had tar-paper on the outside walls and roof and, as at the Sterling, had no water or toilet inside. Those homes with children had bunk beds put along the back wall. As soon as our house was livable we moved in. Gumbo was very bad on the roads here when it rained and we always struck across the field to the railroad track during wet weather, otherwise you could lose your footwear in the gumbo."[15]

19. What value have these shacks?

"If it is the intention of the Department to protect the squatters who are on this land I will have to have an Inspector, immediately before the sale, place a valuation on the various shacks, small houses and dwellings which are located thereon, taking the names of the owners of each shack, and have each owner sign an agreement either agreeing or refusing to accept the valuation placed by the Inspector if the property is to be purchased by some other purchaser other than themselves..."[16]

I can find no record that he ever owned a homestead.

---

20. Insert here a caesura, to indicate a brief silent pause, during which time is not counted.

---

21. Anna quietly sings from the *Dies Irae*, after the two Josephs— her husband and her brother—die within days of one another:

> With Thy sheep a place provide me,
> From the goats afar divide me,
> To Thy right hand do Thou guide me.

(How many horned cattle, horses, sheep, goats? A hundred years later, I suspect none.)

---

22. From the *Dies Irae*, the Lacrimosa: Full of tears will be that day

The Spanish influenza hit Drumheller hard, partly because the living conditions for many were terrible. This was an era where Drumheller was called Hell's Hole: primitive shacks, latrines that were never cleaned, and latrines that were too close to the water supply. At first, I think I have found conflicting reports about hospitals during the flu epidemic. There wasn't one; it wasn't built until 1919. Or there was one. But more detailed reading reveals that one of the doctors had his own small private hospital, and then the school became an emergency hospital, with nurses coming from Calgary and elsewhere to care for the sick. The drugstore owner couldn't get drugs. Neighbours brought soup to those confined to their houses. And you, Joseph Yopek, in your shack with Anna hovering, and by then nine children to care for and protect from infection—did anyone bring you soup? I think you dreamed of a bowl of your mother's barszcz, the beets fermented with rye bread, little cubes of potatoes for bulk, some precious dill. And nearby, did Anna's brother also dream of soup, another version of beets or the česnečka of Moravia, broth pocked with chicken fat and the heads of garlic melted into its goodness as he died alone in his shack excavated in the bank of the Red Deer River? Before both of you were laid in the cemetery on the road to East Coulee, how long were your bodies left in the Miners Hall, newly built and newly requisitioned for a morgue? It was October. Frost at night, the poplars yellow, the potatoes still in the long hills while Anna grieved and tried not to think of the future. Where I waited to untangle this sad story, humming

Stephen Foster's "Hard Times Come Again No More" as I wiped away tears.

———————

23. A nurse sent from Calgary remembered coming to help out during the Spanish flu epidemic:

> "I was up at 7 this morning to look after the sick workmen, and Mike
> Newaz is dying tonight. Poor thing, he's a great big Slav with a mop
> of black hair and he has been suffering agonies... Three women
> died here last night and there are families upon families all laid
> down without any care, but we cannot do any more. I am all right
> so far... (Some have) pneumonia and some dying, and there is no
> place to put them. They screamed, 'Oh, nurse, come and help me.'
> Some of these chaps were down on their knees praying and one poor
> young man about 17 was lying on the floor on a coat with his shoes
> and socks off. I don't think he will live. I sent them a great bottle of
> cough medicine and fever tablets, but I expect they will pass away."

———————

24. I wish to be with you in any way possible. OVID[17]

On my desk, folders and envelopes of papers, some of them in pieces—the remains of my grandmother's life in Alberta, before she met my grandfather, and after. She married him in 1920,

Julia's funeral, 1923

a widow with 8 children and another recently buried in the cemetery where her first husband and her younger brother are also buried. I think of them in their bleak house in Drumheller with its legacy of death and illness—the Spanish flu, diphtheria—and of the graves in the nearby cemetery, the marked ones and the unmarked. In the photographs I've been studying, I have blurry moments when I suspect I'm seeing ghosts. A hat on a chair. A dog watching an empty road, as though in anticipation.

But those ghosts are also my ghosts so it's work I need to do. My grandmother is in my hands, my body, the way I peg sheets to the line on a summer morning, or chop garlic for my own version of česnečka. I am the mother of sons and a daughter who are her great-grandchildren, though they only know her through a couple of photographs, some stories, a long folk song of food they hear when I sing her praises: her soup, her striky, her rich perogies, cabbage rolls tender as butter.

There are ghosts, and ghosts—the magpies we saw in Wayne, as we came out to our car in the early morning from a night in the Rose Deer Hotel, the trees and coarse grasses stiff with frost. Non-migratory, they have been there since the beginning. They might have shadowed my aunts and uncles on their way to school, bread and homemade cheese in a lard bucket. Or watched my grandmother hang out the shreds of laundry on a cold morning in November, the river beginning to freeze over, and the opening of her late brother's house—a dugout in a bank—reminding her daily of his death, as the space in her own bed reminded her of her husband's death, and her youngest daughter who joined them in the cemetery a few months later, holding space for a half-sister still unborn, who would die five years later. The blurry moments in the old photographs: my grandparents' home in Drumheller as a funeral is recorded, a young boy—my father—as he rides his tricycle over the hard earth. If I look long enough, will something be clear? That scratchy signature of my grandmother's first husband on a petition to Ottawa, begging to be allowed to stay in the shack he'd built on land he didn't own. It's all mine, if I can only record it and commemorate it in all its difficult details. If my questions find answers. Even if they don't. I hold the notes of the final prayer, the aria, a psalm my father loved:

Fire, and hail; snow, and vapours; stormy wind fulfilling his word:

Mountains, and all hills; fruitful trees, and all cedars:

Beasts, and all cattle; creeping things, and flying fowl:

Kings of the earth, and all people; princes, and all judges of
the earth:

Both young men, and maidens; old men, and children.

The blurry moment, the blue moment, the morning on the aqua iron bridge over the Rosebud River, just after 7 AM, leaving Wayne and its old buildings, the Last Chance Saloon, abandoned mines, the light on the worn hills opposite, magpies loud in the willows, the moment arriving in my mind as clearly formed as anything: this was part of it. This river, its crossings, memory of dragonflies on its surface, the light on the hills, and the rough song of the magpies. Keep this safe, keep this sacred, I said, and wished I knew the right prayer for such mysteries. In my notebook, I record the light and the smell of water, the flinty Rosebud on its way to the Red Deer, my own heart beating in time to the music of magpies in trees not yet far along enough for leaves. On the roadside, showy milkweed pushing up through the gravel.

I consider the ghost of a child's hand in an ultrasound image, another of a baby's spine, my father on a tricycle, in a little metal car, grey, grey, the propped coffin on a bench in 1923, on the stair by a house I am uncertain is the one my father spent his childhood in or an earlier one that burned. Is it the house built by Joseph Yopek or a later house?

In Drumheller, I thought I'd be able to find my way to the very place where my grandmother stood in the doorway, children around her feet, to look out at her new surroundings, buckets waiting to be filled. (*I'm remembering everything backward.*) I thought I'd know the smell of the earth, the soil that my father ran through in bare feet, and whose dust rose in summer to settle on laundry hung out to dry, on the surfaces of tables by open windows. I thought I'd know it. Recognize the wind. But instead we drove each wide street as though in a foreign country. Following clear and detailed instructions, we missed the turn to the cemetery and had to ask an Asian woman in the uniform of a fast-food outlet, walking home with her head down. She was helpful with her hands, though her English was poor. My grandparents' English was also rudimentary, even after forty years in Canada. How many generations of dragonflies and monarch butterflies? How many generations of children buried among their own babies in the Drumheller cemetery, which we eventually found, decoding the map and the narrow lanes among the dead. How many were buried before the language of grief flowed smooth and clear in the vowels of the new country? Gone, the palatalizations, fricatives, and trills of Central and Eastern Europe, the stops, the lost aspirates. Did I ever speak to my grandparents alone, apart from the urgings of my parents to thank them for gifts or to tell my age yet again, a girl among her brothers, arranged by height, each of us self-conscious in the summer heat, the long drive behind us and the promise of our cousins ahead. The promise of the Exhibition, each with a dollar in our pockets.

And did my grandmother ever tell us of her first husband and the shack he built for her on the banks of the Red Deer River, how he dug a garden in preparation, how he went under the earth for coal, some cold potatoes wrapped in clean cloth, or did this come from an aunt, looking back under poplars in a yard, thinking of the distance a family travels, by water, by rail. As far away as the Carpathian Mountains, so far that some of them die on the way to lives of their own? Two babies buried in the Drumheller cemetery, where we saw showy milkweed, heard the click of beetle wings, the small strophe of local music almost too faint to hear. And a husband, a brother, either sleeping in the mass flu grave or else somewhere forgotten, their own journey abandoned too soon. No, I don't believe she told me. All of this I gleaned from a sentence here or there, a fragment of song, a remembered prayer on a string of beads. Something that happened a century ago, west of the 4th Meridian.

Migratory, like monarchs, we find our own urgent way to a place where the sun and earth greet us, give us rest. We find our place among wild plants on a roadside, we hear beetles and the lazy drone of bees. If we sit on the grass and let the dry wind ruffle our hair, will the voices come to us again?

NOTES

1. The Dominion Land Act of 1871 regulated the surveying of the Canadian prairies. The meridians were used to number the ranges west of Fort Garry, the 1st Meridian, with the 4th Meridian at 110° W marking the border between Saskatchewan and Alberta. The term "west of the 4th Meridian" is used in legal descriptions of sections and quarter-sections of land in Alberta, sometimes written as W4.

2. From "How Do Animals Keep from Getting Lost?" M.R. O'Connor, May 28, 2016, *New Yorker.*

3. The voices in this libretto are as follows: Joseph Yopek, first husband of Anna Kishkan, nee Klusova; Anna; Theresa Kishkan, granddaughter of Anna; Anthony Kishkan, son of Anna and her second husband John Kishkan, father of Theresa; J.G., of the Newcastle Coal Co. Limited; Davidson & McRae, Drumheller Townsite Department; J.F. Drew, Inspector of School Lands; W. R. Cumming, Mayor of Drumheller; Frank ... (unintelligible signature), Department of the Interior (Canada), School Lands Branch; Mrs. Bond, a resident of Drumheller; Frank A. Collins, Superintendent, Department of the Interior (Canada), School Lands Branch; an anonymous nurse from Calgary. From letter accompanying petition to purchase sent by squatters to the Minister of the Interior, May, 1916.

4. From petition to purchase ... portions of the sw1/4 of Section 11, Township 29, Range 20, west of the 4th Meridian, [unintelligible] contiguous to the Town of Drumheller, in the Province of Alberta.

5. A real-estate subsidiary of the Canadian Northern Railway.

6. Klus was the family name. My grandmother's maiden name, Klusova, took the feminine form.

7. From a Memorandum from the Department of the Interior, in response to concerns by the Canadian Northern Railway Company, anxious to have the squatters removed from the land in order to further develop the townsite, and by the Newcastle Coal Company, asking that the lots

be sold to individuals in their employment and needed for the local economy.

8. The questions asked are from the Alberta Genealogical Society's website, www.abgenealogy.ca (Retrieved April, 2016). I quote from documents I discovered in microfilm 2539, file number 31177, held in the Provincial Archives of Alberta. I thank my son, Dr. Forrest Pass, for helping me to decode some of the material from the file and for his ongoing interest in unravelling the mysteries of our family's history.

9. From a letter of January 26, 1915, sent to the Superintendent of School Lands by a representative of the Newcastle Coal Co. Limited, (Sgd) Newcastle Coal Co. Limited.

10. Letter from the Canadian Northern Railway Company, Winnipeg, June 28, 1915. Superintendent of School Lands, City. Davidson & McRae, Townsite Department.

11. The poem by Kenzaburo Oe is from *Death by Water,* Deborah Boliver Boehm, trans. New York: Grove Press, 2015.

12. From a letter written by the Inspector of School Lands, June 14, 1915.

13. From a letter to Joseph and Paul Yopek, September 29, 1914.

14. From a Great North Western Telegram, September 30, 1916, to Deputy Minister of Interior, Ottawa, Ont. From W.R. Cumming, Mayor. Original spacing preserved.

15. The passage about the Bond family home is from *Hills of Home,* by the Drumheller Valley History Association. Drumheller, Alberta: Drumheller Valley History Association, 1973, pp. 124-5.

16. Letter from Department of the Interior Superintendent Frank A. Collins, December 1, 1916. "I beg to point out that on some of these acres there will be from 4 to 8 shacks and if the owners of these shacks agree to accept the price as placed by the Valuator it will practically prevent any other purchaser from buying these portions, as the shacks would possibly be of no value to the purchaser, and if the owners of the shacks do not attend and buy the land I fear the sale of many pieces may be abortive."

17. The epigraph by Ovid is from *Tristia* 5:1, line 80, translated by my daughter Angelica Pass. It's a gift to have her available to ask about Latin and Greek and to locate sources in the complicated corpus of classical texts.

The Poignant house, with Naval communication station towers in background, 1945

POIGNANT MOUNTAIN

I remember that mountain behind the married quarters at Ridgedale, on Matsqui Prairie, remember it for the way it framed our days, its shadows and the sound of wind through its trees, a mountain named for the couple who lived in an old house not far from us, with some shade trees, a creek where my brothers caught frogs. We walked with our mother up the mountain itself for blueberries on summer mornings, lard pails washed out for our bounty. Poinit, I'd heard, in my family's pronunciation. Or maybe Pointit. The name of the couple, Mr. And Mrs. Poin(t)it; and the name of their mountain. By Sumas Mountain, or maybe even a shoulder of it. But I never saw it on any map, so in later years I thought it was perhaps a family name, a familiar name—the way our family calls the big island on Ruby Lake "White Pine," though you won't find that on any map.

But I then found a photograph of the home of Albin Poignant, found during a search through online images at a Fraser Valley museum. The farm you can see to the left is the Sward's, I believe, where my father and I sometimes took a jug for buttermilk (we loved it, cold and sour, with a good dash of salt over top), and those towers are the antennae for telegraph trans-missions at CFB Matsqui where, at three different periods, my father worked for five years when I was young. My first memo-ries spring from our house at the very end of the ten bungalows that housed families of men working at the base—a memory of snow, of a blue bike, of my brother coming home across the field behind the houses with a rocking horse on his back, pulled from a slough where it had been abandoned after being taken from our yard several years earlier. In the memory was recognition: that something I'd once had but hadn't remembered losing was returning to me, a painted wooden horse coming home through tall grass.

On a contemporary map of the area, fifty-five years later, I see Beharrell Road, and I think that was the road where my father tried to teach my mother to drive. It wasn't a success. She knew the basic rules about steering and how to use a clutch. She'd practised with him along the boulevard in front of family housing at the radar base, my brothers and I watching from a safe place, while she ground the gears and reversed in a wobbly line. When they went out for road driving, they had to take us, as we were too young to leave at home on our own. So they warned us to be silent, but it was hard not to scream when my mother drove us into a ditch and my father shouted in exasper-ation. Then my mother wept into her hands over the steering wheel, and my father drove us home.

We knew the Beharrell family, namesakes of the road, who had a farm near Pringle's Store at Ridgedale. We stayed with them once on our way back to Victoria from Halifax where my father had been stationed for two years. I was allowed to ride their horse, and visit their barn with hissing cats protecting nests of tiny kittens. At their huge dining room table, we sat for a dinner of rich food, and I hoped the meal would go on forever. It was a house with dim cool rooms, windows shaded by full trees. My heart explored, found corners I couldn't bear to leave, chairs emptied of generations of inhabitants.

Another farm, on Fore Road, also took me deeply into its fields and rooms, the high reach of its hayloft. The daughters of that farm babysat my brothers and me on the occasional evening when my parents went out, and from that contact, my parents got to know their mother and father: Skinny and Myrtle Gillberg. We bought milk from them, one brother going over on his bike with a gallon jar in the carrier. Their old farmhouse was painted turquoise and had no indoor bathroom, though there was a toilet of sorts in the cellar. We stayed with them sometimes, on trips through the Valley as we headed from a later home in Victoria to our grandparents in Edmonton; and what a decision if I had to pee after dark. To the outhouse behind the house itself, smelling of active bowels and lime, dense spider-webbing in every corner, and loud with flies; or down to the cellar with a candle or a kerosene lantern? A tree by the front door trailed tent caterpillars summer after summer, and if one landed in Mrs. Gillberg's hair, she'd scream. This amused my brothers, but I also hated the feel of them on my face or in my hair. Their whiskery bodies made my skin crawl, and their guts were horrible. There was a couch in the Gillbergs' living room that had survived the flooding of the Fraser River in 1948. I'd press my face to its shaggy arms and imagine I smelled the river. To get to the second storey in that house, you had to creep into a closet in the one bedroom on the main floor, push aside the dresses and winter coats, and find a narrow flight of stairs. The daughters slept up there, and one of them smoked, hanging out the window so her parents wouldn't smell the evidence on her when she came down from her tower, smoothing her hair as she emerged from the closet.

By the road a partly subterranean bunker housed milk cans kept for pickup—this was before milking parlours, before pumps. The bunker was cement and very cool, even on hot summer days. Once my older brothers tricked me into going down into it by myself, later in the day after the milk had been collected, and when I entered, they closed the door, telling me that when they turned the knob a certain way, water would fill the small space. I sat on a bench, waiting for the end. It didn't occur to me that it might be a prank or a joke—I saw nothing funny in the prospect—and I cried softly as I waited, thinking of everything I would miss: my teddy bear Georgie; our dog Star; Christmas morning; the taste of my mother's chocolate chip cookies, warm from the oven. When my brothers finally opened the door and helped me to the surface, they begged me not to tell our parents.

Mrs. Gillberg had soft arms and made the best pancakes I've ever eaten, richly coloured from the eggs we hunted for in the barn or outbuildings, sometimes finding them still warm from the hens. She'd fry bacon cured from the farm's own pigs and let us drink coffee well-diluted with milk. And the milk! The jugs she brought to the table had thick layers of cream on top that I'd scoop out with a finger, if no one was looking. She made pies and biscuits and cakes yellow with eggs and cream. If I'd wondered before where dairy products came from, I knew after one or two meals at her table, having seen the jars of milk come into the house and those eggs on her counter waiting to be wiped clean. It hadn't occurred to me until then to wonder how the hens produced them. And if we timed our visits to the barn just right, we could see Mr. Gillberg or his hired man, Chappy, on three-legged stools, faces against the cows' flanks, squeezing milk from their udders into metal buckets while cats waited in the shadows. On summer mornings, sunlight poured through the open loft doors, where hay bales were thrown from the wagon, fresh from the fields, and swallows entered like swift wishes, dropping feathers that we caught in midair.

I attended a baby shower for someone at the Ridgedale Community Hall with my mother while my older brothers were at school and my younger one played at the home of a neighbour. The women were all dressed in their best, several of them, my mother included, wearing their fur coats. My mother's was muskrat, which made me wonder because sometimes we saw those animals in the creeks leading down from Poignant Mountain and I couldn't imagine how a wild animal could be turned into a coat, with sleeves and deep pockets, lined with silk. The baby shower involved games that the women entered into joyfully, making ribbons from the gift wrappings into hats, and then watching as the mother-to-be held little sweaters and nightgowns to her face before handing them around the circle. Gifts included soft blankets and rattles, diaper pins with clasps of pink and blue, and tiny bonnets of pleated cotton. Tea was served in china cups, and each woman had provided squares or cookies or tarts, arranged on paper doilies spread over delicate plates. Most of the women smoked, holding their cigarettes between fingers that sported neatly shaped and coloured nails, even though many of them were farm wives, their hands moving through the haze of smoke, gesturing and emphatic as they talked and drank their tea. I became drowsy among the murmurs and the smoky laughter, and found the table where the coats had been carefully laid, curling up among the muskrat and camelhair until someone lifted me gently from the pile and said it was time to go home. I found my mother's legs to bury my face in, and rubbed it against the lower edge of her fur coat, then realized that everyone was laughing at me because it wasn't my mother at all—her coat was still folded on the table. My mother explained later that the woman whose legs I'd hugged so ardently was not a mother at all.

My older brothers went to a school in Matsqui, but it had no kindergarten class, so I stayed home, waiting for the bus to return them each day. Some days one of the girls they returned with handed out the crusts of her sandwiches, soft white bread that we never had at home. I'd crouch on the boulevard like a wild child, wolfing down crusts with their faint taste of tuna fish or ham. And what did I do, those long days waiting for my older brothers? I made shelters in tall grass with my younger brother and several other children too young to attend school, rooms we created by lying on the ground and crushing the grass flat, shaping each room to our bodies' dimensions, ceilings the blue sky. We couldn't be seen from the row of houses, though our mothers could hear our voices and knew we were safe from possible danger.

In our airy house, we sometimes exchanged our clothes. Whose idea was it? I don't remember. But we were caught out when I put my undershirt on inside out, and my mother discovered grass seeds in its thin cotton. There was conferring, there were questions: how far had the damage gone? I was used to wearing my older brothers' hand-me-downs so it was no big deal to wear a boy's shorts for an hour, to hand my underpants to another child and wear his or hers for the duration of our game.

But some of the other mothers nipped that game in the bud, their mouths tight and fierce as they led their children home. The grass slowly rose again in our abandoned rooms.

Once or twice that I remember, I waited with my parents for my brothers' release from their classrooms on Friday afternoons, and then we drove over the Mission bridge to shop at Eaton's. We'd eat supper at a cafe on the main street of Mission City, a hamburger or egg salad sandwiches, accompanied by milk-shakes served from tall metal beakers beaded with moisture. The waitress poured some of the contents into a glass and then left the beaker on the counter so we could drink the remainder. We'd suck up every last drop with our straws, running its tip around the bottom of the glass until our father scolded us for making farting sounds with the straw. Stop that, he'd order, the vein on his forehead pulsating in the way it did when he lost his temper. We usually knew just how far we could push him, but sometimes we underestimated and he'd take us outside and wal-lop us on the sidewalk. Our mother never protested. We were wilfulness, to be controlled. We were wildness, to be brought to a civilized state by the rod.

Once my oldest brother climbed Poignant Mountain with a friend, and they failed to come home for dinner. The men of the married quarters went out with flashlights, calling, calling into the darkness. In my bed, I listened, wondering what the boys had wanted, that they would go without permission to sit

Mum and Dan, Gord (on Mum's lap), me, Steve, circa 1958

on a cliff on the mountain, among the blueberry bushes and the sky, and forget to come home. Finally the men found them, and it was as though a stranger came into our house; a boy had gone out as my brother, known and loved, and returned with a secret. Whatever the mountain held in its privacy, my brother had taken in, along with creek water and the smell of the night in his old jacket.

All this time later, I sometimes dream of those years, the radio base, the fields behind it, the rocking horse coming home from the slough, slung over my brother's back. In one dream, we

Postcard, Mission City, 1955

waited at the bridge to Mission City. A boat was passing underneath, and so the bridge came up, slowly; we could hear the
system creaking. Hydraulic? It was late afternoon, perhaps five
PM, and it must have been late fall or early winter because the
river was almost dark. The beehive burners on the other side
glowed. You could tell their dimensions by the lines of fire showing between the joinery. My brothers and I teased and argued on
the back seat while our parents smoked and talked quietly in
the front. In the dream, the car filled with acrid tobacco smoke
and the smell of sulphur. My older brothers, playful, held my
younger brother's head down and farted on him. I laughed, to
protect myself. I knew if I objected, I would also be the victim.

This was later, when I was 14, my father working once more
at the radio base, and we lived again on Matsqui Prairie, though

not in the married quarters. My parents had found an old white farmhouse to rent, its attic claimed by bats. That bridge was a crossing I loved. From Matsqui to Mission, from our side of the river to the other, huge sturgeon lurking in the water. The open prairie to the shadow of the mountains. From the dailiness of our lives—supper at the arborite table, the six of us in our usual places, dog sleeping nearby—to the lights of Mission City. The cafe on Main Street. Sandwiches or burgers on thick white plates, a garnish of pickle, a small mound of rippled potato chips or fries. A milkshake served in a tall glass. We'd park up behind Eaton's. You'd enter the store adjacent to the parking lot, and because the store was built on a hill, what seemed like the ground floor was actually the second. I bought panty hose there, my first pair. And the fabric for the apron I had to make in Home Economics 8. If I walked out on Main Street later, I could see the Star Weekly sign on the store across the street and smell woodsmoke on the wind. Was it from the mills along the river or the Eddy Match Company a little east of the bridge?

All the mystery of waiting at the river for the bridge to come down, the dark water, the glowing of the beehive burners, the anticipation of an egg salad sandwich and a chocolate milkshake. A crossing I loved. From Matsqui to Mission, from our side of the river to the other. And returning, driving home over the bridge again, from the shadow of the mountains to the open prairie, along Riverside Road, past Miss Kemprud's where we went for ice-cream during Sunday drives, past the school my younger brother attended, past the hall where, at the age of five, I'd been in a fashion show—I modelled a tartan skirt and short-sleeved sweater from Eaton's, which I hated and which my mother bought for me afterwards—along Harris Road, then Glenmore, house lights golden in the black fields, turning right on Townshipline Road until we reached our own driveway, the quiet barn with its sleeping cows, and the sound of frogs loud in the slough. I've turned a dream into a memory. But in fact both are the same.

Across dark water, I went from childhood to adolescence. The house at the end of the row by the radio base was not the house we returned to from Mission City in the dream that was also a memory. It was a white house on a farm on a long road, but that road led back to the foot of Poignant Mountain, forgotten and then found, lard pails stained by blueberries and abandoned on the verge, a small girl huddled in the cool bunker where the milk waited to be collected and where I wait now with her for the end of the world.

NOTES

1 This essay is a memory map and remembers its way along roads, up mountains, over rivers, and into towns and villages I knew as a child in the Fraser Valley, first in the early 1960s and then later that decade. To anyone familiar with that landscape and whose own memory differs, I offer this

observation from Herman Melville: "It is not down in any map; true places never are."

2. The photograph of Albin Poignant's house is used with permission, The Reach P3263, Abbotsford Museum.

The old house behind the Mahon house, early 20th century

· · · · · · · · · ·

BALLAST

A book of the inequality in the curve of the sides of ships.

A book of the inequality in the position of the tiller.

A book of the inequality in the keel of ships.

LEONARDO DA VINCI[1]

---

1. In representing wind, besides the bending of the boughs and the reversing of their leaves towards the quarter whence the wind comes, you should also represent them amid clouds of fine dust mingled with the troubled air. (Da Vinci *Notebooks*, 470)

In small communities or the old neighbourhoods of larger ones—towns, cities, even the places where rural areas have been absorbed by suburban sprawl—it's not uncommon to find sturdy plantings that have survived many decades. Lilacs in cold climates often bloom exuberantly without any care at all. They come true from suckers, meaning they haven't been grafted. I saw them in abandoned gardens near Barkerville, in the Yukon, in Drumheller where my grandparents settled: "This plant is native to the rocky slopes of the central Balkans in southeastern Europe in mountainous parts of Romania, Bulgaria, Serbia and Macedonia."[2] Delphiniums send up their tall spires, blue as the sky in these abandoned gardens, though in well-tended borders, they are routinely eaten by slugs. And roses—well, you can tell where homestead gardens were located by the profuse canes of old species ramblers and rugosas—Dr. Van Fleet, Blanc Double de Coubert. Unpruned, unwatered, they cascade over whatever supports might be nearby: a wire fence, a tree (sometimes even a lilac), the remains of a staircase leading nowhere.

I've taken my share of cuttings. My three New Dawn roses come from the garden of my parents' neighbour, Daisy Harknett. In her eighties, she told me how her mother started the roses from a slip given her by the Ferry sisters, a duo who lived nearby in one of the oldest houses in Saanich. The New Dawns, the palest pink (the colour of my baby daughter's shoulders when Daisy gave me these cuttings), tangled themselves in the limbs of an equally ancient pear tree. That tree, with its cargo of roses! Later, the property was subdivided, and the back part, with an old stable, was sold. A man pulled out the rose with a backhoe. I don't know where he took it.

Some old wood, some new wood, said Daisy Harknett. So I cut pieces with both. I dipped the lower part of the wood in rooting hormone (though I could have used a tea of willow bark) and stuck them into little pots of soil. And now my New Dawns tumble over a beam, a pergola, and the front door of my house. The pear tree, with its heavy crop of honeyed fruit, is lost now forever, consigned to the same fire as the rotting fence posts, the stable door. Yet anyone who ate one of those beauties must surely remember the flavour. I took a bag of them to one of my classes at the University of Victoria in 1974 and handed them around to my classmates. The instructor, an Irish poet of some note, ate a couple of the pears in quick succession and said they were the best he'd ever tasted. Years later he published a memoir with ripe pears in the title, and although I haven't read it, not yet, I'd like to think it might be an unconscious homage to Daisy's pears.

In the woods between Elk Lake and Beaver Lake, I remember an abandoned house completely knitted into place by honeysuckle and roses. Knitted into my memory by roses of a kind I've never seen since, apple-scented, white, and humming with bees. On my black horse, I approached with the sense that here was an ancient fairy tale hidden in the woods. Which were not wild exactly but remnant—a few forgotten apple trees, pruned by deer, beaked hazelnut, even laburnum. I entered the tale, as a girl will, with a sense of wonder and expectation. I tied my horse to a tree and tried to peer in the windows laced every which way with canes. And though there might have been a prince sleeping within, he didn't wake. Not then. Not for years. And now he sleeps in our bed, by a window framed with honeysuckle brought by his mother and a trumpet vine given us by her neighbour in Nanaimo, plants held in trust by us, in memory of women gone from this earth.

———————————

2. When the sun is covered by clouds, objects are less conspicuous, because there is little difference between the light and shade of the trees and of the buildings being illuminated by the brightness of the atmosphere which surrounds the objects in such a way that the shadows are few, and these few fade away so that their outline is lost in haze. DA VINCI NOTEBOOKS, 459

Mann Avenue, Royal Oak, Victoria, B.C. It's so different now, that road where my parents bought their first house in 1969—they were in their forties but until then hadn't settled anywhere because of my father's military transfers. You got to our side of Mann Avenue by turning off Glanford Avenue. Then, the road ended at a small clearing by Colquitz Creek, a place where lovers parked on weekend nights and where I rode my horse when school finished on winter afternoons. Another short section of Mann Avenue led off Wilkinson Road. Now the two sections have been connected, and the fields have been filled with townhouses and a subdivision, and only those of us who were young in the 1960s, or earlier, remember the orchard where the Mahon family had grown apples for cider in the early twentieth century and where the remaining son Bill, probably in his late

Mahon House, circa 1940

sixties, lived in the beautiful old house with his bulldog Winnie. Where a man who lived on Christmas Hill rented some of the orchard for his young cattle. Where I'd retreat sometimes when our house got too noisy and where I'd read under apple trees and where once a calf grazed around me as though I didn't exist, so still was I in the soft grass.

Some weekends, young men came to visit. When the garbage men collected on Mann Avenue on Monday morning, the sound of crashing glass rang through the air as they emptied Bill's can into the truck. Gin bottles, my mother would giggle. Thereafter, I always associated gin with young men coming for the weekend. I only entered the house once, when I knocked on the door to sell some forgotten item for a school fund-raiser. The house smelled old. Old in itself, inhabited by old residents: Bill and Winnie. I waited inside, in the vestibule, by the front door, while Bill searched for his wallet. He wore slippers, I remember, and he shuffled and trembled, as though he'd been drinking. When the young men came, you never saw a sign of life for the entire weekend. They pulled the blinds and put Winnie out by herself to pee in the yard and then sleep on the path leading to the front door.

Behind Bill's house, but still on his property, sat another even older house. Daisy Harknett said it was the oldest house in Saanich. Her son Carl rented it for a time, with his wife and young child. Occasionally I babysat for them, walking down the rough lane behind Bill's grand house to the shabby wooden building behind while Winnie barked from the window. It was a little scary to be in the house, alone, or at least the only one awake, while a small child slept upstairs and the trees creaked outside in the wind. Walking back to my house past midnight was like walking into another century, the one I knew.

In the past, I've asked archivists for information on the street where I lived as a teenager. Specifically, I had questions about two houses that I thought had historical value, in that I believed they might have been designed and built by Bert Footner, who'd also built many of the bungalows in the orchard community of Walhachin, on the Thompson River between Cache Creek and Kamloops, around 1909-1910. After a long career build-ing houses and bridges in the Sudan and California, and failing as a farmer, Mr. Footner retired to live on our street with his wife and unmarried daughter; two houses adjacent to his prop-erty had the lines of the colonial bungalows he'd designed at Walhachin, as did the house he lived in, now long demolished, though the other two still exist. Michael Kluckner wrote about those bungalows on his Vanishing B.C. website:

> Footner's houses reflected the colonial bungalow form that had evolved from indigenous Indian architecture and become the stan-dard country house in England's hot-climate colonies, especially Australia. A high hipped roof (sometimes built as a true pyramid) provided a natural insulator to keep the main-floor rooms cool. Win-dows were set to catch views and breezes. Porches, sometimes open and wrapping around the house, other times screened, provided extra heat shelter.[3]

I asked the archivists long after the Mahon houses had been torn down and tract houses erected in their lost orchard, but one archivist insisted the road hadn't existed prior to the 1950s, so the houses I was remembering as "old" couldn't possibly have existed either. He was sure of this. Yet the Ferry sisters, who'd lived on the corner of Glanford and Vanalman Avenue in a low Victorian cottage surrounded by wild fields, told my mother that they'd attended dances with other local farm families before WWII and that they'd walked to dances at the Community Hall on West Saanich Road, built by the Quicks. They wore their rubber boots on a trail to Mann Avenue and carried their dancing shoes in a bag along the dark street to Wilkinson Road and over to the bright hall. Neither of them married, but my mother said they had been beauties. I wish I'd paid more attention. My mother had a rose rooted for her by the older of the Ferry sisters, a bright pink climber, and I wish I'd dug it up when my parents sold their house. Somewhere, forgotten now, I found two images of the Mahon houses, so I know they existed, I didn't dream them, and Mann Avenue passed by the front door of the big house where Winnie slept on the path, in sunlight and under clouds, and where the garbage cans waited with gin bottles hidden inside. Mr. Footner came out to watch me ride my horse Marlin down Mann Avenue, asking me to stop for a moment so he could pat Marlin's neck. A fine animal, he assured me as he leaned on his shooting stick, his daughter Mollie standing beside him.

Everyone I knew on that street is gone now. My parents. Bill Mahon. Daisy Harknett, who grew those wonderful pears and whose mother always said that you needed some old wood and new wood to make a cutting take. Mine took, all three of them, and I have large tangly New Dawn roses to thank her for all these years later. The Footners—Mr. and Mrs. and their grown daughter, Mollie. The neighbours on the other side, the

Cromacks, gone too. I was visiting my parents when the husband passed away. From the open window of the guest room on that side of the house, I heard his grown children cry as they stood in the back yard and planned what to do next.

Yet I carry the neighbourhood in my memory, sometimes an awkward weight and sometimes a richly textured landscape. To every house since, the ballast of those years, heavy in certain chambers of my heart, though I would be bereft if the heaviness was absent. Every spring, the New Dawns bloom, and I remember the branches of the pear tree that held them in light.

---

3. Young plants have more transparent leaves and a more lustrous bark than old ones; and particularly the walnut is lighter coloured in May than in September. DA VINCI NOTEBOOKS, 421

What I took to Ottawa in May when we went to visit our older son, Forrest, his wife, Manon, and their little son, Arthur: seeds (rucola del orto and lettuce leaf basil from the Home Hardware on Commercial Drive), little transplants of volunteer kale—two kinds—wrapped in damp paper towel and tucked into a Zip-lock bag. An eye-watering piece of horseradish root. A linen cloth. Blackberry jam.

What I brought home: 6 tomatillo volunteers from Forrest and Manon's compost, perhaps an inch high, with tiny threads of root trailing from the stems. I wrapped them in damp paper towel and put them in the same Ziplock bag the kale had travelled in, along with a dozen cilantro seedlings, equally frail. Each seedling went into a pot of good soil. Two months later, the tomatillo plants were seven feet high, laden with fruit. All winter we open jars of roasted green salsa flavoured with garden onions, peppers, and lemon juice from our Meyer tree, and brightened with handfuls of that cilantro that had thrived too in the summer's heat. Each jar a memory of a family gathering in Ottawa, all of us (for the others had also come to Ottawa, Brendan, Cristen, and Kelly from Edmonton, and Angelica from Victoria) at the table eating prime rib roast from the Weatherall Brothers butcher shop around the corner, garnished with the horseradish, its root already tucked into a corner by the canoe. In our compost, the overripe tomatillos are sleeping the long winter months away and will hopefully sprout to carry on the memory of their Ottawa Valley origins.

What I also took to Ottawa that May: a rooted sucker from the wisteria growing by our west-facing deck. It was, in turn, once a sucker dug up from the woodshed where long adventurous strands of our original wisteria formed roots along their lengths. The one growing up one of the woodshed's posts was carried along a beam, cut from the big cedar we had taken down years before and which contained pumpkin seeds deep in the heartwood, encloses the patio. That wisteria came from John's mother who brought rooted cuttings of wisteria, two kinds of honeysuckle—the Japanese *Lonicera japonica* 'Halliana' and the beautiful Dutch *Lonicera periclymenum* 'Serotina'—and other ornamentals that bloom every year. She's been dead since 2012, but her plants help us to remember her generosity. She carried rooted shoots of the original family wisteria in turn from her mother's garden in Suffolk, wrapped in damp paper in her suitcase after one of her annual summer visits to her mum. Have you anything to declare, I imagine her being asked, and like me (who carries acorns and interesting cones and seeds from everywhere I visit), she took a deep breath, keeping inside every important reason for children to continue their parents' gardens, and said no. In her suitcase, the long roots of her mother's mint, the perennial geraniums.

What I took home: the memory of all of them laughing, baby Kelly crawling on the grass, the sound of glasses clinking, the excitement of waking in the morning with the knowledge that I could be among them for another two days, another day, a few more hours. A few more hours.

What we took to Edmonton, first time we visited, by car, in October: a tea set—cups, saucers, cream and sugar, side plates— that had travelled with John's grandmother the first time she came to Canada, a gift for John's parents to celebrate a wedding anniversary; a silver teapot from my parents' home; some bedding from my parents, including the Hudson Bay blanket they'd bought themselves as fortieth wedding anniversary gift and which they used for camping trips; a wooden salad bowl and servers; silver soup spoons I'd found in the thrift shop, just before we'd bought others with a set from a junk store in Falkland. We took a case of wine as a housewarming gift, lugged up the elevator to the high suite overlooking the Saskatchewan River, wines chosen on warm days wandering through the south Okanagan Valley where we'd camped with our children twenty-five years earlier.

What we took home: the close harmony of their life together; laughter at their table, set for Thanksgiving with old plates and silver spoons; the memory of a picnic at Elk Island Park where the huge bodies of the plains bison scented the air and the larches were just beginning to turn.

---

4. Every shoot and every fruit is produced above the insertion [in the axil] of its leaf which serves it as a mother, giving it water from the rain and moisture from the dew which falls at night from above, and often it protects them against the too great heat of the rays of the sun. DA VINCI NOTEBOOKS, 419

This morning I was thinking about ballast, its weight and permanence, and was in the process of wondering, not aloud exactly but certainly on the page, about a rose I have growing over a railing on the west-facing deck, our garden "room" if we have one: it's the deck where our table resides in summer, where we eat our dinners, sit with guests at night with glasses of single malt, listening for loons down on Sakinaw Lake.

The rose came from one of the annual spring plant sales at the Community Hall when we first lived here; you brought your box with you, and you got there early because everyone wanted the tomatoes or irises or Muriel Cameron's dahlia tubers or bits of Vi Tyner's roses. I'm not sure this one came from Vi Tyner, who did give me moss roses, a soft pink one and another one deeper pink in colour. But it grows everywhere—old home-steads, seaside gardens, along fences in semi-industrial areas as if remembering a former house, ancient care. It grows across from the Post Office in Madeira Park, for example, and I don't know if it ever gets pruned or watered. And there's a place on the highway, near Middlepoint, where one grew for years and years, until it was absorbed by the forest taking over the site of a cabin that I believed burned to the ground before we arrived in 1981.

I'd thought a little about trying to identify it but somehow never did. And somehow today was the day, so I took my rose encyclopedia and a cup of coffee out to the table and went through, page by page. Until I came to 'American Pillar.' Bred by Dr. Van Fleet in 1902. A very prolific and widespread rose, and yes, it will survive any kind of neglect, it seems.

I'm interested in how plants travel, how they are carried to new places, how they are botanical palimpsests, in a way. And how they hold stories, some plain and true, and some cryptic. In Placentia, Newfoundland, two autumns ago, we stayed in a beau-tiful old Second Empire bed-and-breakfast inn, with a lovely garden in front, overlooking the gut or channel connecting two arms of water. And a little photo essay in the entrance hall detailed the restoration work done on the house, adding that the old roses in the garden had come in soil serving as ships' ballast, the ship having come from Ireland. Imagine the seeds, damp in a hold, salt-stung, coming to live on the other side of the world on the Avalon Peninsula. What did they carry of their origins in Ireland, how long did the soil sit in a boat in that Placentia gut, who carried it out of the hold to land, and who noticed the roses growing from the heap?

I have roses from gardens no longer extant. Vi Tyner's for example, which provided the moss roses as well as white violets, yellow flag irises, and a root of *Viburnum opulus* that promptly died, though I was always too timid to tell her. A 'Tuscany Superb' that came from a cutting given me at a birthday party held perhaps 30 years ago in a house that has long been torn down.

---

5. That part of the body will be most illuminated which is hit by the luminous ray coming between right angles. DA VINCI NOTE-BOOKS, 420

In the thatched house at the Ukrainian Cultural Village Museum near Edmonton, some rough linens are on display, lengths of bright woven cloth on the benches of the good room where visitors would be brought. It was the room where a wedding was celebrated by seventy guests eating and dancing, the "owner" told us, a man from Bukovina whose daughter-in-law worked in the fenced garden. There were potatoes, beets, feathery fronds of dill everywhere, self-sown, a hardy variety: did it travel with the family from Bukovina, a twist of paper containing its seeds, its beloved flavour, the flavour of home? Along with Black Krim tomatoes, Koda cabbages (from Polish relations), the Viktoria Ukrainskaya peas? Seeds traded with Mennonites for their own hoarded heritage, with Sudeten Germans and Croatians and Armenians for cucumbers, along with Lyaliuk from Belarus. We ate cabbage rolls and cucumber salad green with dill at the snack bar, and I tasted my way back to the long table set up in the sunlit backyard of my aunt and uncle where my father's family gathered every time we visited Edmonton, the women in the kitchen all morning rolling dough and filling pedeha with soft mashed potato and cheese curd and sliced green onions so strong my eyes watered. Slices of dry sausage dark with caraway and rolls with hard crusts. My uncles held a fist of bread and a glass full of something clear that they drank down, grimaced, then laughed. We had our own drink, raspberry juice with a whiff of vinegar, compot it was called, and was poured from quart jars, murky with floating fruit, that we were asked to bring from a certain shelf in the cellar where spiderwebs draped the windows. As if descending to the hold of a ship, we crept down stairs steep as a ladder, to the sway of ancient shelves and their cargo of bottled fruit, beets, the precious compot. Sometimes we pretended to be the uncles, drinking deeply and dancing with our glasses raised high, laughing and slurring our speech. We rolled on the sunny grass like drunks fallen under the weight of what was held in the glass, plums distilled into memories of the old country, trees around the houses their parents had been born in, had left, had never returned to, the passage of ships across water almost a dream. We didn't know what sorrows they carried in their pockets, hidden away at times like those, but tolerated by wives who cooked and wiped at red faces with a tea towel damp with steam.

---

6. A place is most luminous when it is most remote from mountains.
DA VINCI NOTEBOOKS, 464

Among my father's effects was an old Avon box with papers and a few photographs, all unsorted. I'd never seen these things until after my parents died, and yet they have become my ballast as I enter my sixties, wanting to know the places left behind by my grandparents and what their early lives in Canada were like. An archive of the deep past containing faces like my own, languages I'll never speak, a memory of rain on a tin roof in a shack in Beverly where my grandparents lived after leaving Drumheller, lilacs against the porch.

A photograph of Johann Kiszkan as a young man, dark-eyed and stern. His booklet from his time in *Infanterie regiment Freiherr von Reinländer* No. 24, 6. Kompagnie, with stamps and signatures. His travel papers, issued from Vienna in the name of Franz Joseph I, allowing him to come to Canada but not to return to Bukovina, and used as "Exhibit A," translated by Eva [surname unclear. Szkavorok?], before a Commissioner for Oaths in and for Alberta, 25th day of April A.D. 1949. A photograph of two young women, one of them resembling him. A photograph of Anna Klusova and her new husband, Josef Yopek. Foothills of the Carpathians at dawn. The wooded Beskydys, small wooden house at the foot, tucked into an orchard of plum trees. A rosary. A postcard in Polish.

NOTES

1. The epigraph as well as the passages threaded throughout this essay are from *The Notebooks of Leonardo Da Vinci*, compiled and edited from the original manuscripts by Jean Paul Richter, New York: Dover Publications, 1970. This is an unabridged edition of *The Literary Words of Leonardo Da Vinci*, first published in London in 1883.

2. The quotation on lilacs is from http://plantwatch.naturealberta.ca/choose-your-plants/common-purple-lilac/ (Retrieved May, 2016).

3. Michael Kluckner's website: https://www.michaelkluckner.com/bciw6wal-hachin.html (Retrieved May, 2016).

$a, b \in \mathbb{Z}$

$b + a = q_0 b$
$r_0 + r_1 + r_1 < r_0$
$r_0 = q_{1} r_1 + r_2, \quad r_2 < r_1$

...

(1) $\qquad r_{N-1} = q_{N+1} r_N + r_{N+1}, \quad r_{N+1} < r_N$
(2) $\qquad r_N = q_{N+2} r_{N+1}$

$\implies r_{N+1}$ divides r_N by (2).
$\implies r_{N+1}$ divides r_{N-1} by (1).
...
...
$\implies r_{N+1}$ divides a and b.

Any divisor of a, b divides $r_0 = a - q_0 b$.
$\implies$ any divisor of a, b divides $r_1 = b - q_1 r_0$.
...
...
...
$\implies$ any divisor of a, b divides r_{N+1}.

$\implies gcd(a, b) = r_{N+1}$.

Melba apple tree in winter, with algorithm

· · · · · · · · · ·

EUCLID'S ORCHARD[1]

Come now as the sun goes down.

See how evening greens the grass.

Is it not as though we had already gathered it

And saved it up inside us...

RILKE[2]

I never imagined we'd abandon our orchard. It was one of the first things my husband and I planned as we worked on the house where we hoped to raise our children and write our books.

We spent a year or two learning the lay of our land—where the sun rose, where it set, and when; where the warm hollows were, where we might sit and look out to blossoming trees in spring. John dreamed of walking under their gracious boughs, and I hoped for a wide gate so we could back our pickup truck right into the orchard to load up the fruit. Dreams of space and abundance, the scent of apple blossom and then the jars of preserves on the pantry shelves to take us through winter. In anticipation, I bought a big willow basket to hold the apples of the future.

We hired a backhoe to clear the bottom land at the foot of the bluffs where our house stood. A half acre, perhaps, a gentle slope, fringed with alders and a few cedars. Rocky, but with pockets of soil—and anyway, we reasoned, we could build up the planting areas with manure, seaweed, and compost.

Those first trees—bought from a man who collected heritage varieties and had an apple tasting weekend at his orchard on Norwest Bay Road in West Sechelt: we tasted, then ordered a Melba, a Golden Nugget, a Cox's Orange Pippin. The trees were tiny, and we planted them reverently, shrouding them in old gillnet salvaged from the dump. Later we bought a Transparent, too, for early pies and sauce. Pears: a winter variety, and one for eating out of hand in late summer. Two cherries, a golden plum, a greengage, an Italian prune. Hazelnuts. Each tree had a cage made with sticks and gillnet. But eventually we fenced the whole area, a long process, which required that some postholes be created by building up cairns of rock and then cementing the posts into the stone. We used green wire. We used chicken wire. We used strands of wire connected to a battery in order to give any animal trying to enter the orchard an electrical reminder not to trespass. The truth of our location, though—below a mountain, in woods threaded with game trails—slowly dawned on us.

It wasn't until our big dog Lily came to us, in her second year, after her original family moved elsewhere, that a combination of fence and canine reliably kept deer and bears from the trees.

The deer liked them any time of year. The tiny buds and leaves, then the fruit. The bears liked to drag down boughs and tear fruit from them, leaving huge piles of golden scats everywhere. Even the grouse liked to climb out to the ends of the branches and nip at the buds and green fruit. Eventually the Roosevelt elk discovered the trees. But Lily slept with both ears attentive to the slightest noise from the orchard, and often at night we'd hear her barking, hear the barking grow fainter as she pursued intruders deep into the woods. Sometimes she slept in the grass under the trees. She took her job seriously.

On the transitive property of equality: *Things that are equal to the same thing are also equal to one another.* EUCLID'S 1ST *AXIOM*[3]

I don't remember when coyotes arrived at the north end of the Sechelt Peninsula where we built our house in the early 1980s. There were coyotes in Vancouver, certainly, but not on our part of the coast. People said that a healthy cougar population kept them away. And in those years, we heard regular reports of cougars passing through schoolyards, hovering on the beaches, prowling the hiking trails along the spine of the peninsula. We heard one once when we were living in a tent with our first baby, building during the day and collapsing onto the foam mattress within the canvas walls at night. We heard it scream— yes, it *was* blood-curdling—and our dog, Friday, a big English sheepdog cross who slept under the tarp by the entrance to the tent, began to dig under the plywood platform we'd built so that the tent would be off the damp ground; she wanted the safety of the small space between earth and plywood. Now I know the cougar was probably a female in estrus, alerting males to her availability. But in the darkness, it was as wild a sound as I've ever heard.

Coyotes were spotted south of us, in Gibsons, perhaps in the early 1990s. How did they arrive? Our peninsula is connected to the mainland by mountains, too arduous for road travel; and people arrive here by ferry, crossing Howe Sound from Horseshoe Bay, a distance too far for a coyote to swim. But perhaps a pair, a pregnant female, even a family intrepidly heading out for fresh territory, could find a way through the mountains. After the initial sightings in Gibsons, we began to hear reports of coyotes mid-Coast, then at Halfmoon Bay, and finally we began to see them occasionally where we live near the northern end of the peninsula.

During the summer of 2005, we drove back and forth from the little village near us, over three evenings, for a music festival we were involved with, and each night, coming home around 11, we saw a pup in exactly the same place by the Malaspina Ranch, hovering in the area where coltsfoot and wild hazelnuts grow. In our car headlights, it was pale, alert, and obviously waiting for a signal from nearby parents to let it know it was safe to return to whatever it was doing. Which might have been eating stranded fish in the muddy bay below the old barn across the road.

And driving back from dinner with friends late one night, we saw a coyote family feasting on a deer recently hit by a car on a wide part of the highway. We stopped to watch, and the adults backed away, but only a short distance, their teeth bared, not willing to risk us taking the carcass from their young. It was abundance to be defended. If we'd opened the car doors and walked towards them, I think we'd have regretted it. It was a window into a wild moment, mirroring our own care and concern for our young. Who were grown up by then and living far away.

––––––––––

Things which coincide with one another are equal to one another.
EUCLID'S 4TH *AXIOM*

All my life, I've felt at home in language. I began to read when I was five, before I began primary school at six. Fifty-two years later, I still remember how I would rush to my room, having just read a library book in grade two, and try to write one myself. The stories involved horses, perfect families, ranches tucked into box canyons, or else camping trips on the edge of rivers or beaches resembling the ones my own family camped near during those years. When I got older, I still tried to write stories, spinning out the intricate webs of words until I ran out of them, unable to finish. When my children were small, I loved to gather them around me so we could read a book together. One sustaining memory has us immersed in Arthur Ransome's *Winter Holiday* over the Christmas season when we had an unexpected flurry of snow. That story tells of a group of children quarantined from school because one of them has mumps. The lake they live by has frozen, so they spend the days skating and traveling its length on an improvised ice sledge. On our own holiday on another continent, many decades later, we remembered the book as we skated another frozen lake. The Ransome book was also about childhood languages—the children use Morse code and other kinds of signals to communicate—and about an igloo (that spiral of snow blocks), an expedition to an imaginary Arctic, camping on an ice-bound houseboat, and the dire rescue of a sheep. My children hauled out our collection of sleds and tried to recreate the book's adventures. Lit by the warmth of a good story, they used its codes to make their own, in action if not in words.

We understood one another in those days. The books, the movies we watched together, the challenges of locating constellations in the story-laden summer skies. It never occurred to me that other vocabularies would enter our lexicon and that I would comprehend fewer and fewer of the codes my children used to navigate their way through the world.

Especially my younger son, Brendan, who, from the time he entered adolescence, became more and more fluent in the language of mathematics. His high-school teachers praised his abilities, and his exam scores were often perfect. The school directed him towards the sciences, in part because they had strong teaching in that area and in part because no one thought the humanities were truly an option. Except his parents, both writers with English degrees. When Brendan went to university, we advised him to take a range of courses in his first year so he could find out if he really wanted to study physics and math or if he might veer toward history, perhaps theatre. Math and physics it was though, and when he went on to do graduate work in Toronto, he concentrated on math. I tried to understand the nuances of a partial differential equation but felt as though I were reading Sanskrit.

The boy who once counted stars as he reclined on the dry moss of our orchard, windfallen apples in hand as he translated the density of the Milky Way into individual points of light, who walked the alder path with his grandfather and explained his theory of negative numbers as the dry leaves rustled at their feet, began to travel to conferences all over North America and Europe. I was so proud of him for doing well in his chosen field, but the field existed in a dimension foreign to me. Could I touch its ground? Was a line something I could trace my finger along, ease my needle along with its tail of cotton thread? And did the line end or keep going to infinity?

To produce a finite straight line continuously in a straight line. EUCLID'S
2ND *POSTULATE*

I've made dozens of quilts over nearly 30 years. I begin each
in a kind of heat of creation, ideas and colors clamoring in my
imagination, my hands almost shaking with the urgency to sew.
And yet, I am anything but skillful with the tools of quilting.
My own are pretty low-tech—scissors, a cutting mat and wheel,
packets of quilting needles, three wooden hoops of varying sizes,
and a frame of PVC pipe with clamps to fix a section of quilt
into a taut surface for stitching. I have other bits and pieces I
almost never use—a thimble, a leather finger protector, books of
patterns that are somehow never the patterns I want to make. I
invent my own, badly, and scribble them onto scraps of paper—
old manuscript pages, envelopes, the pieces of light card that
come in pantyhose.

If equals are added to equals, the sums are equal. EUCLID'S 2ND *AXIOM*

In mathematics, Euclid's orchard is an array of one-dimensional
"trees" of unit height planted at the lattice points in one quadrant
of a square lattice, the square lattice being a type of lattice found
in two-dimensional Euclidian space. It's taken me most of the
fall to grasp what this means.

And then it turns out I've imagined it as something far more complicated than it is. After asking Brendan about the difference between the one- and two-dimensional spaces, he sends me an email two sentences long:

> I think two-dimensional just refers to the lattice that the orchard lives on (like in a real orchard, where the surface of the ground is two-dimensional).
> One-dimensional refers to the trees themselves (which extend vertically from the two-dimensional ground).

Euclid didn't know the Earth curved, that it was a globe loose in the universe. That lying on the grass in our orchard, we were turning as the Earth turned, around our own dear sun. The warmth of it on our faces, the bees alive in the apple blossom, the drift of violets in the dry grass, a woodpecker drumming in the woods nearby—all this was momentary as we moved towards night. But we could anticipate the beauty of it again and again, light bending around a star.

I want to find a way to sew the graphic representation of Euclid's orchard. For years I've quilted to find a way to explore texture, to try to bring a one-dimensional space into something resembling two: runnels of cotton pierced with small stitches; stars slightly embossed on a bed of dark velvet. I think of these surfaces as a kind of landscape relief, something for the hands to read as though reaching down from the sky to learn something about the Earth—hills of brown corduroy, cabins built of strips of coloured fabrics, a red hearth at each center; an indigo river alive with the ghosts of salmon, their eggs glittering among the stones below them.

I want to recreate those neglected trees rising from the surface of the ground, the clovers and wiry grasses around their trunks, the spring daffodils I planted in hope, and all the teeming biota under the earth: worms tunneling through the porous soil, the burrows of field mice, the root systems of the native wildflowers. I want the nematodes, the protozoa, fungi, the broken rocks, minerals, the decaying organic matter creating the humus needed to keep the soil healthy. I want to commemorate the dream of an ever-thriving orchard, alive in our stories when we now gather together ("Do you remember the time we slept on the moss?" "Remember walking up the driveway and surprising a herd of elk that winter, snapping off the lower boughs").

Some mornings in late spring, I walked down the driveway and saw the lattice of chicken wire strung with dew-covered spiderwebs, glistening in the sun. The chicken wire was composed of cells like the hexagonal cells bees create from wax taken from abdominal glands of the worker bees and softened in the mouths of house bees. The wax is shaped by their bodies into circles, which quickly form into hexagons. Mathematicians such as Pappus of Alexandria[4] attributed this to "a certain geometrical forethought" on the part of the bees—"Bees, then, know just this fact which is useful to them, that the hexagon is greater than the square and the triangle and will hold more honey for the same expenditure of material in constructing each"—but modern physics suggests that it is less calculated. Bees make cells that are circular in cross-section, all packed together like bubbles. Surface tension in the soft wax pulls the cell walls into hexagonal, threefold junctions as the wax hardens. I think I am with Pappus on this one, though. Try dropping circles of wax as close to one another as possible onto a flat surface and watch what happens. I've done this, several times, dropping wax from a beeswax candle (for authenticity), and the results are *not* hexagonal threefold junctions.

The pear blossoms smelled of sandalwood. Pressing our faces to branches heavy with flowers, we'd breathe in small insects, brush pollen from our noses.

On early summer mornings, our orchard was like a corner of a Botticelli painting, tiny violets and daisies and wild roses, heart's-ease and self-heal, wands of Columbia tiger lilies, four or six blooms to a stalk. To paint them, to have some long longed-for gift to capture each in its specificity, the opening sequences timed for bees, other pollinators: the art of utility yoked to beauty.

We added strands of barbed wire between 4-by-4 posts after deer kept getting in between the green wire and electric. But the deer persisted. The bears, too.

There was that gate, generously wide so we could bring in our truck with that big willow basket, for when the trees produced the harvests we thought might be possible. I filed recipes for apple preserves, for plum jams, for bottled cherries in exotic liqueurs. One year I scavenged enough pears to process in Mason jars, and we had pies of Transparent apples encased in buttery pastry. Some years this happened, that we managed to harvest enough for ourselves against the constant predations of bears.

When Lily barked at dawn or dusk, when we could hear animals crashing in the woods near the orchard, or within it, we'd go down with sticks and pans to make enough noise to raise the dead. Or scare off the living.

Finally we gave up, though the chicken wire and the green-covered wire still surround the trees, gaping in places where the animals muscled their way through. We returned the battery, almost unused, because the bears didn't care about electric fences any more than they were afraid of barbed wire. They didn't give a fig—and when they discovered the fig tree by the house, they ravaged the young fruit until the tree grew too tall for them to reach, and because the trunk grew against the side of the house, they didn't care to climb it. But down in the orchard, they'd usually eat every pear on the trees and then shit out golden pulp as their own parting barb.

That all right angles are equal to one another. EUCLID'S 4TH POSTULATE

I've been making quilts for over two and a half decades, a quarter century, a process expressed as a series of fractions, reduced to the lowest common denominator: hours of the day stretching into weeks, then years. A basket of cloth, a bedcover, a bag of scraps.

I realized soon into the world of quilts that my carelessness with measurements and my slap-dash way of planning would cause me grief. "All right angles are equal," said Euclid. I assume that cutting those angles precisely would mean that a quilt would become a series of beautiful and elegant geometries. Instead, my stars, some of them the Variable star, composed of triangles within squares, aiming for yellow constellations in dark skies, are lopsided, the imprecise points of stars forced into clumsy arrangements. Another style of star, the Ohio, usually turns out even worse. With every new quilt, I promise myself I will slow down, take my time—the time needed, each step conducted with care and attention to detail, no matter how small and seemingly insignificant. And with each new quilt, I am again in the middle of a chaotic equation (though not differential), cutting at random, my notes scribbled on the back of an old envelope. How many triangles will I need to make these stars? How many strips of light cotton, then dark, to build the log cabins, each with a tiny red square dead centre: the hearth?

We built our house wall by wall, framing each wall on the floor platform, then lifting them into place. And the wood stove blazes in the middle of the kitchen, drawing us to its heat, to sit on chairs with quilts on the backs for extra warmth.

The cleared area where we planted our orchard was not geometrical. Or wait, maybe it was. *Is*, because it's still there, though the alders grow closer, the limbs of cedars, once trimmed, lean over the collapsing fences, the salal and huckleberry reclaim what was truthfully theirs. From the air, would it resemble a trapezoid? I try to envision it, the points that might be considered corners. Angles. Is it faintly scalene? I realize I know far too little about everything. Math. Quilts. Coyotes. The requirements of fruit trees. The complicated tangle of inherited traits.

I thought it was too late to learn something about mathematics. All my adult life, I've had dreams in which I am about to take an exam in math, an important exam, but I can't find the classroom. Or else I realize I haven't even attended the classes all term, all year, a number of years. In the dreams, I recriminate myself, walk towards the exam heavy with dread, knowing that I will never graduate because I can't even begin to do the math. I wake, feeling hopeless.

Then Brendan suggested I take a course. Or at least sign out the Great Courses offering, Joy of Math, taught by Arthur Benjamin, from the local library.[5]

Turns out Benjamin's a good teacher, a magician, an international performer of mental calculation. Mostly what intrigues me about him is that he took the time to deconstruct simple mathematical principals and to show the fledgling student (I was 59 when I watched the 24 lectures on my computer screen) the complex structures in a way that highlights their beauty and interconnectedness. Arithmetic to prime numbers to combinatorics. At the heart of those questions he poses, you can figure out your possible outfits if you have three pairs of shoes, four skirts, seven blouses, and two cardigans. And maybe two pairs of jeans. A very useful skill for packing a single carry-on suitcase for extended travel... Prime numbers. Fibonacci numbers to algebra, which finally made a kind of sense as I watched. Proofs to geometry, which I've always sort of liked because of how quilts are so often based on geometric forms within blocks, but made not by pencil on paper: rather by hands sewing cotton, easing corners together, working out how to represent a tulip with rectangles and triangles. And then pi. Beautiful pi: π. I wanted to think of it as a pie, literally, because that was something I knew about, could bake if the bears left enough apples, hold a knife over top and plan the cuts to allow the greatest number of people to receive a more or less equal slice. Pi always made me tremble in my boots, but how reassuring to learn that it's a constant, a ratio (a circle's circumference to its diameter), and not something wildly arcane or terrifying.

One day, while taking a break from the Joy of Math, I was reading the *Vancouver Sun* obituaries on a quest to find six degrees of separation in at least one of the entries, when I saw the obituary for my eighth-grade math teacher. (*Things which coincide with one another...*) He was a charming man but shares responsibility (with me, of course) for derailing me from any kind of deep interest or understanding of mathematics. He had a wandering eye and assigned the prettiest girls in the class—I wasn't one of them—to the front row, nearest his desk and the blackboard. He flirted with them so emphatically, leaning over their desks to explain the problem on the board so quietly that the rest of us knew we were doomed. At noon he drove around in his convertible with the top down, and an array of lovely young girls arranged on the seats like beauty queens. He taught me something important, but it wasn't math.

The degrees of separation became the possibility of connection. That summer of 2014, I felt confident enough to try to grasp some new ideas, to understand in particular how every linear transformation can be represented by a matrix, and every matrix corresponds to a unique linear transformation. My transformation might be in language—a new language—and maybe an expansion of my heart and spirit. My feet, long planted firmly in mud and compost and the piles of manure I'd have delivered every few years, might find themselves venturing onto some other pathway. And it would help me to understand how an orchard, and the legacies of genetics, could be given a new context in a quilt. I was thinking of how to turn my attention, newly attuned to mathematic pattern, to fabric—its dimensions, its willingness to be shaped and manipulated into something at once practical and encoded. An equation for love and for the years devoted to growing trees, sons, a daughter, and then understanding the point at which one had to simply let them go.

———————

One autumn evening, under brilliant stars, a white coyote crossed the highway as we drove home from Oyster Bay. Its eyes glowed, and its ears were beautifully shaped, like receptors—every sound of the night entering them: owls, mice skittering under dry grass, a raccoon leading her kits to eat apples in moonlight, even the skeins of snow geese heading south in the darkness, muttering and calling, their navigational system a form of quantum entanglement.

———————

I am interested in mathematics only as a creative art. G.H. HARDY

I ordered a copy of G.H. Hardy's *A Mathematician's Apology*[6] in a moment of nervous energy, thinking: I will continue trying to learn about the history if not the practice of math. And it turns out I didn't need to be nervous. Hardy was a mathematician who wrote elegantly and clearly about youth, the beauty of pure math, and who said, late in life, that his biggest contribution to his field was his mentorship of the Indian mathematician, Srinivasa Ramanujan. His book is an essay, really, a reflection on having given his life to the pursuit of something for which he was not interested in extending a practical application. "It is a melancholy experience for a professional mathematician to find himself writing about mathematics. The function of a mathematician is to do something, to prove new theorems, to add to mathematics, and not to talk about what he or other mathematicians have done."

I read it with a kind of surprised delight. His mind constantly made connections between math and music, math and poetry. He was interested in pattern. And I realize, with some surprise, that this is why I take such pleasure in following his sentences—each well-constructed and leading the reader logically through his thinking.

Pattern has been an abiding concern of mine, too, one I am constantly looking for ways to explore. In language, in my orchard, in fabric as I imagine ways to build a whole of small fragments of cotton. I don't use formal patterns but try to let colours of the materials and the ideas I am interested in determine how I will cut and piece together and shape into something practical.

This is where I part company with Hardy. I want my quilts to have a function, perhaps because I am a woman and for thousands of years women have used the necessary materials of daily life to create art. Subversive art, in many ways. How could a table cover or a bed cover or a basket for collecting seeds and roots be considered art at all? And therein lies the subversion. Women, some believe, created string by twining short sections of stem and root and fibre together. And string functioned as yarn, as thread, as anything that might be twisted and woven and made into something durable. If that was your job and you were a woman, then you found ways to make it beautiful. You dyed and compounded pigments, you wove, you knit, you wrapped your children and yourself in lengths of fabric you might have stamped or printed or appliquéd, and you grew food, cooked it, served it, and it was beautiful, though no one called it art. The women, though, knew what it felt like to have our hands immersed in rich wool or finding our way through a complex schema of linen thread, of silk yarn, of pounded nettle fibre, with a whole palette of grasses and reeds for imbrications.

I am driven to make a quilt using the patterns I'm finding in mathematics. I can't sew well enough to piece together De Finetti's triangle, but I am thinking about ways to explore these ideas. To look at them for their graphic interest, and behind that, the subversive notions they reveal to me about genetics, parenthood, the correspondences between life and science and art, and to try to draw together the beautiful and the durable. Or to track the grid of trees planted for optimum light and air, but as lovely as a trellis in a medieval manuscript, holding a vine close to a stone wall for its warmth, but making an intricate lattice-work of branch and leaf.

In retrospect, I wonder how we knew the coyotes had arrived in our area to stay when we saw them so infrequently. Other big animals occurred ephemerally in the forests and mountains surrounding us. Cougars passed through. Wolves too—tailing the herds of Roosevelt elk reintroduced in the early 1980s in part to try an innovative way of maintaining the brush growing under the big Cheekeye-Dunsmuir power line.

We knew about the coyotes because they left scats on our driveway, in the hollows of moss in the orchard, on the nearby trails we hiked regularly, and even along the highway we walked to collect our mail at the community boxes about half a mile away. Every time we walked, we saw the scats. If we were on a trail, the scats were in the middle. The animals wanted anyone using the trail to know they'd been there. On the edges of the highway—a sign that the animals had mastered the knowledge of traffic—the piles were right on the human-worn margins.

And they were—*are*—fascinating. Coyotes are ominivores. They eat rodents, frogs and other amphibians (but not toads because their skins are bitter), reptiles, fish, crustaceans, birds, larger mammals that they can either kill or scavenge, grass (which helps them to digest fur and bones, I've read, and which also serves to scour parasites from their intestines), birdseed, and all kinds of fruit and vegetables. Once we watched a young pup hold salal branches down with its foot so it could reach the ripening berries, plucking them delicately one at a time. We've noticed more fur and bones in spring, when rodent populations are highest. And sometimes the scats seem to be composed entirely of grass. Once, the head and neck of a garter snake, scales still intact. Bloody flesh gives them a darker color. Fruit—crabapples, wild cherries, even elderberries–give them bulk. Seeds and fur make them grey. And if they're lucky enough to find a source of dry dog or cat food, the scats resemble those of canines.

Even though they were mostly invisible, we knew they were around and felt lucky when we saw them. Luckier still when we heard them. We live far from the nearest village and can usually hear emergency vehicles coming from a distance. But if there are coyotes in the immediate vicinity, they begin to howl before we hear the sirens, and by the time the ambulance or police car is near our house, on its way to the ferry or to deal with a collision on the highway below us, there's a cacophony of siren and coyote accompaniment. A wild orchestration for voices and synthesizer—longitudinal waves coming toward us, bending and refracting the long length of the highway. Sound nowhere and everywhere.

The Dualities: A meditation on correspondence

I read Edward Frenkel's *Love and Math: The Heart of Hidden Reality*,[7] entranced by the ideas of a unified theory linking math and physics because it gave me courage to think that other things could be connected if one could only find the right language, the right equation. I have no idea what automorphic functions are, and when I try to figure them out, I quietly give up. But not before I find sites online that offer—again—the most beautiful graphics, with colors and feathery patterns and I want so much to make something of this. Not mathematics, but something resembling how I feel when I look at its vocabularies, its imagery.

Frenkel's book is filled with anecdotes—partly a memoir of his relationship with physics and math as well as with significant mathematicians, living and dead (and truly the dead ones are as lively as the ones still with us). The book contains recipes for borscht, memories of conversations, jokes, deeply serious analyses of current research and future possibilities, and even an account of the genesis and making of his film, *Rites of Love and Math*, in which a mathematician discovers a formula for love. He quotes from Henry David Thoreau: "The most distinct and beautiful statement of any truth must take at last the mathematical form. We might so simplify the rules of moral philosophy, as well as of arithmetic, that one formula would express them both." And I thought of Euclid again: *Things which coincide with one another are equal to one another.* Though I suspect my own apprehension of an equals sign is not necessarily the same as a mathematician's. But in my reading, I've come across the work of the 16th century Welsh physician and mathematician Robert Record who is credited with introducing algebra to England and also with first using the equals sign in his book, *The Whetstone of Witte*: "To avoid the tedious repetition of these words: 'is equal to,' I will set (as I do often in work use) a pair of parallels, or Gemowe lines, of one length (thus =), because no two things can be more equal."[8]

Of course, I thought. That makes such sense: parallels.

You can draw a straight line between any two points. EUCLID'S 1ST POSTULATE

For the longest time, I couldn't think how to represent the ideas that catch my attention. Not wholly, because although I know, for instance, that an algebraic equation is a combination of numbers and letters equivalent to a sentence in language, I read those sentences as clumsily as a child might try to sound out a simple passage from a primary reader. In grade one, I was praised for being able to decode the word "something," a word I'd not yet encountered in my library books or volumes borrowed from my older brothers' shelves. The sentence, in a Dick and Jane reader, was "Mother makes something." In our reading circle, in the front classroom of the Annex at Sir James Douglas Elementary School on Fairfield Road in Victoria, I sat in one of the two plaid dresses my father bought me on a naval trip to Hong Kong and used my finger to keep my place not just on the page but in each letter of each word, symbols becoming meaning in the chalky air of that classroom with its generous windows looking out on

Garry oaks and houses on the lower slopes of Moss Rocks. I never had that deep illumination with the math lectures, though I enjoyed brief moments when I realized what relationship was at the heart of a differential equation or how the Fibonacci sequence worked.

But some of the graphics are beautiful. An arrangement of stars and squares illustrating dominant and recessive phenotypes. An ancestry tree showing that the reproductive numbers of drone and worker bees in any generation follows the Fibonacci sequence. I choose 12 of my favorites and then spend more time figuring out how to reproduce them on fabric. Should I appliqué? Should I use colored pens on plain cotton? Should I free-associate and simply create abstract versions of each concept? Try to reproduce a grid echoing our original orchard? None of these are quite right.

Then I remember that an ink-jet printer will print on cloth, and after tracking down a source of specially prepared cloth (which seems more practical than trying to pass regular cloth, stiffened with freezer paper, through a printer), I buy some.

I decide to use a cotton print to border each block—it is a little like the illustration of Mendel's theory of how hereditary characteristic are passed from parent to offspring. I decide this will be a quilt to hang on a wall. The treated cloth is stiff, and I don't think it will drape softly enough for a bed cover.

Making this quilt is like the long process towards learning even the most basic concepts in mathematics. I sit with the cloth.

I look at the printer. I wonder. I ponder. My heart starts to race. I hear the voice I've heard all my life when something new presents itself. Who do you think you are?

———————

One day a single light brown coyote came out of the woods and walked by my window. It had all the time in the world. It passed the wing of rooms where my children grew up. It passed the windows they looked out at night, first thing in the morning, drawing their curtains to let sunlight in or the grey light of winter, in excitement, lonely or sleepless, in good health and bad, dazzled with new love or sorrow, at the lack of it, on the eve of their birthdays, new ventures, on the eve of leaving home. I went to the back of the house to see where the animal was headed, but it did what coyotes do, a trick I wish I could also learn. It dematerialized. Vanished into thin air.

———————

I'll use red thread for this quilt, this assemblage of geometric investigations, algorithms, spirals, and ratios, an orchard's grid in plain and printed cotton, small stitches to draw layer to layer, capillaries to help the blood of our relationship circulate through the images and actual fabric of my thinking. Red thread, long strands carried by the needles I will prepare, three at a time, to allow me to push and pull the red lengths in and out, to meditate between the past and present, to contemplate the future, to secure with tiny knots the end of each fragment of thought.

I love looking at Pascal's triangle. I don't understand a lot of the language used to talk about it, but the way I understand it, simply, is this: each number is the sum of the two directly above it, or below it, depending on how you are viewing the triangle. In a way, it's like genetics, except that contained in the numbers (those binomial coefficients) are other sums: the grandparents, the great-grandparents, and those falling back into history, into prehistory, but who might emerge in the integer at the top of the triangle in some form undreamed of by the two below (I am looking at the triangle from the bottom up...).

This is the way I'd like genetics to work, in a way. Two numbers adding up to one, a system that is predictable and easy to plot on six rows of Pascal's triangle. But of course human beings are nowhere as tidy as numbers, though we can chart the DNA sequences within the 23 chromosome pairs in the cell nuclei and within mitochondria.

A mathematician, like a painter or a poet, is a maker of patterns. If his patterns are more permanent than theirs, it is because they are made with ideas. G.H. HARDY

At my desk, I look up to see two large brindled coyotes lope out of the bush and across the grass in front of my study. In the past, I've heard coyotes in the woods just south of our house and suspect there's a den there used year after year. Once, reading in bed late at night, my husband and I heard a pair mating— the rhythmic grunts and growls, the high-pitched squeals, a passionate duet, tempo changing until all we could hear was an urgent expressive finale, and then silence. Though running, these two also seemed at ease in their surroundings, coming out of the woods where there's a rough game trail used by deer and elk, and crossing the grass as though they'd done it many times before, on their way to the orchard. I called my husband to see, but by the time we opened the back door, they'd disappeared.

––––––––––

The stray, the unexpected variable.

One apple tree remains under my care. It's a Merton Beauty, bought as a tiny plant at a produce store in Sechelt. An organic gardener had grafted interesting varieties to dwarf rootstock, and I chose one almost at random. Merton Beauty is a cross between Ellison's Orange and Cox's Orange Pippin. For years, ours sat sort of sullenly in a little circle of stones near the garden shed, caged in chicken wire. I'd water it, give it the occasional mulch of compost and drink of fish emulsion. A few frail blossoms, an inch or two of new growth. Then it produced some fruit that was delicious. The information I've read about this variety stresses the aromatic flavor of the apples—their spicy taste, redolent of pears, cinnamon, aniseed. I can't say I noticed those particular notes, but the skins were pretty, russeted at the shoulders, and the flesh was crisp, with a true flavor of apple. Not the empty watery taste of many supermarket apples, sprayed, waxed, gassed, and stored for months.

When we rebuilt the vegetable garden after the septic field over which the garden was first made needed repairs, I replanted the Merton Beauty within the newly fenced area. I gave it lots of mushroom manure, bone meal, alfalfa pellets, and a long drink of liquid kelp to help it settle into its new home, a raised bed I called Apple Round.

We also have five crabapple trees up near the house. Two of them, growing in tandem, were given us twenty-five years or more ago by John's mother, and each spring they bloom like debutantes in deep pink gowns. Working near them, we hear the bees. Most autumns a bear comes for their scabby fruit that is the size of plums. Beside them I planted a lanky crab from my friend Harold Rhenisch; he told me he'd brought it back from Bella Coola. And further down the driveway are two small crabapples, white-blossomed, with tiny apples the size of cranberries. These came from Roberts Creek. Once upon a time, I made jelly using a combination of crabapples from all five trees, but no one in our house really liked it, and there are so many more rewarding preserves to make in fall so the bears are welcome, if they would only not break branches in their eagerness to gather fruit from the high limbs. And grouse, too, like to graze on the frostbitten apples in late fall. More than once, we've joked about a Thanksgiving dinner of apple-fed grouse, but neither of us has the heart (or gun) to make this happen.

So one eating apple and its array of pollinators. And now a stray. Just beyond the sliding doors that lead from our kitchen to the sundeck, coming up from rocky ground, is a small tree that has revealed itself to be an apple. Not a Pacific crabapple—our native *Malus* (or sometimes *Pyrus*) *fusca*—which is what I thought it was when I finally recognized its leaves and bark. I left it to grow up beyond the pink rambling rose tangled among the deck railings so we could enjoy its blossoms in spring. Last year it had fruit, and they weren't crabs but fairly large green apples: there were four of them, and when it seemed they might be ripe, when they came easily off the branch when twisted a little, I picked one to try it. Not delicious, not even remotely. I think now of Euclid: *The whole is greater than the part.* A tree's beauty is more than the taste of its fruit. But the question of course is how the tree got there. I know that apples don't come true from seed. Blossom from a Merton Beauty, say, is pollinated by an insect bearing reciprocal pollen from another apple—here, it would be a crabapple—and although the resulting apples would be true to their tree, their seeds would be the children of the Merton Beauty and the crabapple. One in ten thousand of those seeds might produce something worth eating. Who are the parents of this stray apple tree? It started growing before the Merton Beauty began its small production of fruit. Did this tree sprout from a seed spit over the side of the deck or excreted by birds or even seeds from the compost into which I regularly deposited cores and peelings from apples given us by friends in autumn? Belle of Boskoops from Joe and Solveigh, for instance, which make delectable fall desserts and cook up into beautiful chutney. Or else a seed from the few rotten apples from the bottom of a box bought from the Hilltop Farm in Spences Bridge, their flavor so intense you could taste dry air, the Thompson River, the minerals drawn up from the soil, faintly redolent of *Artemesia frigida*. This stray is all the more wonderful for its mysterious provenance, its unknown parents, and its uncertain future, for it grows out of a rock cleft, on a dry western slope. I won't dig it up since I have no doubt its roots are anchored in that rock, but I will try to remember to water it occasionally and maybe throw a shovel of manure its way this spring.

The Fibonacci numbers are everywhere in nature, as a numbering system that allows a pinecone or sunflower to pack as many seeds as possible in limited space, to allow cellular information to flow in an efficient way in organisms ranging from the ovaries of the tiniest fish to the fiddleheads of ferns and the leaves of a tightly wrapped head of cabbage. Some days I wander the garden as though through an archive, looking at flower shapes and seed pods to find this miraculous system. In my study, I have pine cones on sills and shelves, gathered from beloved places—the Nicola Valley, the side of the highway in Marble Canyon where a fascinating research project is underway in Pavilion Lake, looking at fossil microbialities, some of the earliest remnants of life on Earth, and from a shelf just north of Lytton overlooking the Thompson River. And a new place too, the Dominion Arboretum in Ottawa where I imagine walking with a future grandchild, sharing my love of trees.

"By dividing a circle into Golden proportions, where the ratio of the arc length are equal to the Golden Ratio, we find the angle of the arcs to be 137.5 degrees. In fact, this is the angle at which adjacent leaves are positioned around the stem."

Phyllotaxis is the term Swiss botanist Charles Bonnet introduced in 1754 for the study of the order of the position of leaves on a stem, how the spiral arrangement allows for optimum exposure to sunlight.[9] It's a term that sinks so naturally into my own lexicon that, looking at the way the leaves are arranged on a dogwood tree near the orchard gate, I think of my children, my brothers, our parents and grandparents, and all the generations of the spiral configured on our own family tree. We are a case study in phyllotaxis, all of us absorbing the light, all of us contributing (*The whole is greater than the part*), even in death, to the ongoing life and vitality of the tree. Though by now, who knows its genus, its specific name.

Braid groups, harmonic analysis: *The whole is greater than the part.*
EUCLID'S 5TH *AXIOM*

A midsummer evening, clear moonlight. Down in the orchard, the coyotes have gone under the fence with their young. How many? I've seen one, heard several others. I've imagined them on the soft grass, tumbling like my children used to play, rolling down the slope over tiny sweet wild strawberries, over the heart-shaped violet leaves, the deep pockets of moss, while around them snakes hid under the lupines. But now in the quiet, I am shaken out of my dreaming because a coyote is singing a long, low passage. A lump forms in my throat as I look out into the night, the sky dusty with stars, a three-quarter moon hanging so perfect over the hidden lake that I think of a stage set, an arranged scene created by strings and wishful thinking. A jagged line of dark horizon and the vertical trees, the line of them rising, then descending as the bar changes, a page of music, the arpeggiated chords, the implied bass line. A pause, a comma of silence. Another coyote joins in, then at least two more. It's a part-song, a madrigal. Each voice is on pitch, but one is low, another high, and several braid themselves in and around the melody line.

See, see, mine own sweet jewel,
See what I have here for my darling:
A robin-redbreast and a starling.
These I give both, in hope to move thee—
And yet thou say'st I do not love thee.[10]

What feast have the parents provided—a flying squirrel, a clutch of frogs, robin nestlings fallen from a tree, a cat from the summer neighbors sound asleep in their beds? *See what I have here for my darling*—I hear the *riso* in the father's line, his extravagant vibrato; and then the *sospiro*—*in hope to move thee*, as the mother nudges the twitching body towards her eager pups. For she knows, oh, she knows, that by summer's end, her young will have gone their own way, far from the natal den in the woods just south of the orchard, forgetting the braided perfection of the family body and its unravelling, the strands unplucked and loose, and *yet thou say'st I do not love thee.*

———————

Twelve quilt blocks wait for me to find an ideal pattern for them. I arrange them on Moravian blueprint, somehow expecting to see logic at work. Do I begin with the first idea I had—the representation of Euclid's orchard, the set of line segments like a trellis to hold the heavy blossoms of fruit trees in May? Or do I find a way to let the blocks tell a story all their own, dense with figurative language? Does it matter?

Blossoms ignite on the long, unpruned branches of the stray apple. The bees are in heaven, their faces buried in the open flowers, rising on legs heavy with pollen to find another, and another. Nearby a sapsucker tests the cotoneaster where the young are brought, year after year, to learn to feed on insects their parents have trapped in pools of sap. Leaning over the railings, I try to see the pattern of the leaves on their stems, because it's a wonder the tree is where it is, rooted in a cleft of rock, its branches nudging into light. It's a wonder, how far children travel from a house buffeted by winter storms, spring rain, the sound of loons nesting summer after summer on the lake just below the forest, and for a time, the promise of fruit from trees planted in their infancy until the orchard was abandoned to the alders and bears, and to the late-coming coyotes who made their home in its remains.

It is the star to every wandering bark
whose worth's unknown although his height be taken.
SHAKESPEARE[11]

Late in my father's life (green-hazel eyes, light brown hair, a sturdy build, a temperament shadowed by melancholia: I have inherited the last two), he talked about reconciling numbers. I believe he had a form of dementia, and this was what he felt called to do, I suppose, though his relationships with his children suffered and could have used the same attention. He hadn't used this term before, or at least not in my memory of him. But he'd been good at statistics and good at mental calculations. After he retired, he sat in a big armchair in one corner of his living room and taught himself celestial navigation. He had a book from Goodwill in which he inscribed his own name— Anthony John Kishkan—and a lifetime of buried ability to learn, a mind that might have been nimble if it had been let free to roam and develop. And maybe it had always been free, maybe I never noticed. (He was the grandfather who walked with Brendan under the alders, discussing negative numbers, the possibilities of zero.) Maybe it was freedom to sit in the big chair with columns of numbers, reconciling them—I have no idea where they came from and why he felt he needed to do this. He used a mechanical pencil, always used one—for sums, for crosswords: "What's a Greek letter used in math to mean a small positive quality?" I wouldn't know so never answered. And he'd repeat the question, querulously. "A Greek letter used, oh never mind, I'll look it up myself." It was the same chair where he sat fifteen years before, newly liberated from his job as a radar technician, and made himself simple tools—a cottage cheese lid cut into a circle and rigged with glass and a tiny mirror became a sextant; cardboard, string, a plastic straw, and a fishing weight became a quadrant. He had patience for this intricate work, but I don't believe he ever did anything beyond finding latitude in his backyard and filling paper with sums. Maybe on the long sea voyages that took him away from us for two or three months at a time—once, six months—to the Orient, Australia, around South America. Maybe he was the sailor who left his bunk and looked at stars at night and wanted to know how to find his way, though by day he worked with radar systems, repairing them, fine-tuning them so the vessels were anything but dependent on celestial navigation. It would have made sense to have learned then, when he could perhaps have applied the knowledge to the dark skies near the Antipodes or approaching Madagascar. But he waited until the early 1980s, after retirement, to sit and work with angles, degrees that became, using the same notebook of graph paper and his mechanical pencil, reconciliations. And I wish now that I'd asked him to show me what a disk of plastic fitted with mirrors could tell a person about where they were on the planet so I could imagine him now, at sea, finding horizon.

––––––––––––

To describe a circle with any center and distance. EUCLID'S 3RD POSTULATE

I tried hard to understand the Joy of Mathematics and realized that I couldn't, except in the broadest possible way. That at the heart of it is an attempt to relate concepts that might not readily suggest themselves to be connected. Number theory and harmonic analysis, for example. And I can only think of those by relating them to the figurative language I learned as a student of literature. Language departing from its logical usage to urge the reader to emotional and intellectual discovery. On that midsummer night, listening to coyotes sing madrigals in our abandoned orchard, I should have remembered Theseus in Shakespeare's *A Midsummer Night's Dream:*

> The poet's eye, in a fine frenzy rolling,
> Doth glance from heaven to earth, from earth to heaven;
> And, as imagination bodies forth,
> The forms of things unknown, the poet's pen
> Turns them to shapes, and gives to airy nothing
> A local habitation and a name.[12]

They were our names, our bodies under the heavens, all of us singing together in different voices to tell the story of our orchard, our time here in this place we have inhabited since— for John and me—1981, and the only way to shape the story is through connotation, not ordinary discourse, though I praise the literal, the specific, but by reaching up into the starlight to parse what lies behind it.

I've tried to puzzle through equations: the arrows, the lines and diacritics, the glyphs, the beautiful characters that look like Greek (*are* Greek)—a notation, a way of assigning symbolic value to constants, function, variables. A way of talking about equalities between variables. It's the chicken-and-egg argument written in the ancient markings of Simonides in wax. Would math work in Chinese characters or the syllabics of the far north? Would flowers still smell sweet if their seed patterns were random? Would a baby ever be born in our extended family without the blue eyes or sturdy legs of a potato-farming ancestor near the Carpathian mountains? Would it matter?

Inside I am stitching a spiral into the layers of the orchard I have pieced together, a snail shell curled into itself. That's what I'll see when I've finished. I begin the spiral at its very heart, keeping my course as even as I can as it opens out and widens. Not the complicated pathways of the sunflower, some turning left, some right, so that an optimal number of seeds are packed in uniformly, or Romanesco broccoli, its arcs within radii resulting in something so intricately beautiful I wonder how anyone could cut into it to eat it. On windowsills, pinecones. The plump Ponderosas, brought home from the Nicola Valley, and a few long Monticolas. They're dry, open, but at the base, where their stalk connected them to their trees, two spirals are still visible, like a relaxed embrace, lovers asleep. My spirals are simple, my hands sewing to follow a path from its knotted source, around and around, until I've learned that my pleasure comes from the journey itself, a needle leading me outward, towards completion. A quilt elegant and sturdy, a sequence emptied of its numbers.

And listen: the coyotes are singing, the deep voice of the father, the rather more shrill voice of the mother—anxious that all her offspring eat well and learn to hunt, to care for their safety in the forest beyond the orchard—and the lilting joyous youngsters unaware that a life is anything other than the moment in moonlight, fresh meat in their stomachs, the old trees with a few apples and pears too small and green for any living thing to be interested in this early in the season.

NOTES

1. In my attempt to understand something of mathematics' beautiful structures, I was assisted and encouraged by my son, Dr. Brendan Pass. This essay is for him, with love.

2. The epigraph is from "The Apple Orchard," by Rainer Maria Rilke, Joanna Macy and Anita Burrows (trans.), in *A Year with Rilke: Daily Readings from the Best of Rainer Maria Rilke*. New York: Harper Collins, 2009.

3. The translations of Euclid's axioms and postulates, used as epigraphs for some of the sections, are from Sir Thomas Heath's translation of *Euclid's Elements*. Cambridge: Cambridge University Press, 1908.

4. Pappus of Alexandria is remembered for his Συναγωγή (*Synagoge*, or *Collection*), composed around AD 340 in 8 books or rolls (of papyrus). Much of the work has been lost, though some exists in fragments translated into Arabic and Latin. His conjecture about honeycombs appears in Book V.

5. At Brendan's suggestion, I spent time with The Joy of Mathematics, one of the Great Courses, taught by Arthur T. Benjamin. I probably understood about 30% of the material presented, but I found it a strangely exhilarating experience. Looking at the offerings on the Great Courses website (www.thegreatcourses.com), I realize I could spent the rest of my life engaged in

online learning. Everything I didn't learn in high school and university or couldn't learn (math, organic chemistry...), or wished I could somehow fit into a course load, now available at my desk, at my fingertips! We'll see.

6. I loved G.H. Hardy's elegant essay, *A Mathematician's Apology*, with a foreword by C.P. Snow. Cambridge: Cambridge University Press, 2002. "A mathematician, like a painter or a poet, is a maker of patterns." And I would add to that list, "a quilter."

7. I thank my friends Joe and Solveigh Harrison for the birthday gift of Edward Frenkel's *Love and Math: The Heart of Hidden Reality*. New York: Basic Books, 2013.

8. I was surprised to find myself reading and enjoying Robert Record's *The Whetstone of Witte* (first published in 1557 and available as a download on several internet sites), once I'd puzzled through its font. I learned that the language and notation of mathematics have gone through many stages of evolution, and this added both to the mystery and the pleasure.

9. The definition of phyllotaxis is from http://jwilson.coe.uga.edu/emat6680/parveen/fib_nature.htm (Retrieved September, 2015).

10. "See, see, mine own sweet jewel" is a madrigal by 16th century composer Thomas Morley.

11. This epigraph is from "Sonnet 116."

12. *A Midsummer Night's Dream* V.i.12-17.

ACKNOWLEDGEMENTS

Family is both the source and inspiration for the essays collected in *Euclid's Orchard*. My immediate and extended family makes my life worth living, and I thank them for their humour, their assistance with everything from historical practice to ancient languages to mathematics to poetry, and their love. The shadowy family I found in old photographs, crumbling certificates, telegrams, documents almost too faded to read, and mass cards is one I am learning to accommodate. During a recent health crisis, they hovered in my imagination and my heart, asking me to record what I could of the faint traces their lives left in Europe and in Canada. I know there's still more to do, and I hope they'll be patient.

I am grateful to Mother Tongue Publishing for its commitment to books that might not otherwise be published, this one included. Publisher Mona Fertig is eager and enthusiastic, and I couldn't ask for a better team than the one she provided: Pearl Luke, for superb editorial guidance; Setareh Ashrafologhalai, for elegant intuitive design; Judith Brand, for a careful copyedit.

I'm delighted that *Euclid's Orchard* is set in Amethyst, a font designed by the important Canadian designer Jim Rimmer (1934–2010), proprietor of Pie Tree Press in New Westminster B.C. He gave the first edition of Amethyst to Mother Tongue in 1998 and taught Peter Haase of Mother Tongue, letterpress printing.

The essay "Euclid's Orchard" was included in *Rooted: The Best New Arboreal Nonfiction*, edited by Josh MacIvor-Andersen, published in the spring of 2017 by Outpost 19.

I thank the Canada Council for the Arts for generous and timely support.

My husband, John Pass, deserves every kind of gratitude. "And listen: the coyotes are singing."

Theresa Kishkan is the author of 12 books of poetry, fiction and literary non-fiction, most recently the novellas *Patrin* (Mother Tongue Publishing, 2015) and *Winter Wren* (Fish Gotta Swim Editions, 2016). Her books have been nominated for many prizes, including the Pushcart Prize (twice), the Ethel Wilson Fiction Prize, the Hubert Evans Non-Fiction Prize (twice), several National Magazine Awards, and the ReLit Award. *Phantom Limb* (Thistledown, 2007) received the Canadian Creative Non-Fiction Collective's inaugural Readers' Choice Award, and "Arbutus menziesii: Makeup Secrets of the Byzantine Madonnas" won the Edna Staebler Personal Essay Contest in 2010. The French translation of *Patrin*, titled *Courtepointe*, will be released by Marchand des feuilles in 2018. She lives with her husband John Pass on the Sechelt Peninsula.

Photo by John Pass